SpringerBriefs in Materials

The SpringerBriefs Series in Materials presents highly relevant, concise monographs on a wide range of topics covering fundamental advances and new applications in the field. Areas of interest include topical information on innovative, structural and functional materials and composites as well as fundamental principles, physical properties, materials theory and design.SpringerBriefs present succinct summaries of cutting-edge research and practical applications across a wide spectrum of fields. Featuring compact volumes of 50 to 125 pages, the series covers a range of content from professional to academic. Typical topics might include:

- A timely report of state-of-the art analytical techniques
- A bridge between new research results, as published in journal articles, and a contextual literature review
- A snapshot of a hot or emerging topic
- An in-depth case study or clinical example
- A presentation of core concepts that students must understand in order to make independent contributions

Briefs are characterized by fast, global electronic dissemination, standard publishing contracts, standardized manuscript preparation and formatting guidelines, and expedited production schedules.

More information about this series at http://www.springer.com/series/10111

Jean-Claude Tedenac

Multicomponent Silicides for Thermoelectric Materials

Phase Stabilities, Synthesis, and Device Tailoring

 Springer

Jean-Claude Tedenac
Université de Montpellier
Montpellier, France

ITMO University
Saint Petersburg, Russia

ISSN 2192-1091 ISSN 2192-1105 (electronic)
SpringerBriefs in Materials
ISBN 978-3-319-58267-2 ISBN 978-3-319-58268-9 (eBook)
DOI 10.1007/978-3-319-58268-9

Library of Congress Control Number: 2017939595

Printed on acid-free paper

This Springer imprint is published by Springer Nature
The registered company is Springer International Publishing AG
The registered company address is: Gewerbestrasse 11, 6330 Cham, Switzerland

Acknowledgements

Appreciated fruitful discussions were made with some colleagues about this subject: Dr. Catherine Colinet from Grenoble-Alpes University, Dr. Alexandre Berche from the University of Montpellier, and Prof. Dr. Lev P. Bulat from ITMO University of Saint Petersburg (Russia).

Contents

Abbreviations

ΔfH	the enthalpy of formation of a compound.
$°G^\theta i$	the Gibbs energy per mole of component i in a phase θ.
μ_i	the chemical potential of component i.
a_i	the activity of component i.
E	the Fermi level value for a given carrier concentration.
Ec	the conduction band edge.
Ev	the valence band edge.
G	the total Gibbs energy of a system.
Gm	the Gibbs energy per mole of components of a system.
$G^\theta i$	the partial Gibbs energy per mole of component i in a phase θ.
$G^\theta m$	the Gibbs energy per mole of components of a phase θ.
N	the total number of moles in the system.
N_A	the Avogadro number with the value of $6.023\ 10^{23}$ atoms per mole.
Ni	the number of moles of the component i.
p	the pressure.
Q	the heat quantity.
R	the gas constant with the value of 8.31451 Jmol^{-1} K^{-1}.
T	the absolute or thermodynamic temperature.
V	the volume.
V.E.C.	the valence electron counts per number of transition element.
x_i	the mole fraction of component i, $x_i = Ni/N$.
y_i	the fraction of constituent i. The sum of constituent i equal 1.
ZT	thermoelectric figure of merit.
ZTmax	the maximum thermoelectric figure of merit.

Chapter 1
Abstract

The thermoelectric efficiency of a material is measured through the value of the so-called ZT (power factor which is defined in the refences [1, 2]). High ZT values in thermoelectric systems are obtained by a good knowledge of the materials involved in their fabrication. In this volume, we will summarize the thermoelectric properties of materials based on silicon compounds and those presently under study. We will show that the phase transformations and phase stabilities of new materials are still unknown, and it is a problem for the synthesis and use of the thermoelectric modules. Applying the modern concepts of thermodynamics, a study of the systems is presented taking as objective their use in materials. We will discuss the development of thermodynamic databases of such practical materials and show the importance of the microstructural evolution of the materials during processing and services for improving their knowledge. For this reason, in this volume, we will present some peculiarities of these materials in light of a phase diagram analysis and show finally how the CALPHAD method is necessary to study such multicomponent materials and to determine the best synthesis process (single crystal growth, powder metallurgy, etc.) [3].

© The Author(s) 2017
J.-C. Tedenac, *Multicomponent Silicides for Thermoelectric Materials*,
SpringerBriefs in Materials, DOI 10.1007/978-3-319-58268-9_1

Chapter 2
Introduction

A review of the state of the art in the research and technology of silicon compounds for thermoelectric applications is presented in this book. It focuses on an aspect often forgotten in materials sciences: the chemical thermodynamics of materials. The silicides are interesting materials for wide applications such as sensors and thermoelectric generators and a lot of specific applications, magnetism for example, when the compounds contain magnetic properties.

Environmentally friendly materials and low cost will be the main attractivity for the mass dissemination of thermoelectric applications; mechanical strengthening and chemical stabilities versus temperature are interesting features in order to get a long-term use, and they are made with abundant elements in the Earth's crust; it is the case of silicides.

Besides the military and space applications and due to the characteristics we underlined, these materials can be applied in civil opportunities and participate to the fight against the global warming in the world.

It will become increasingly effective in encouraging consumers to purchase fuel-efficient vehicles and for manufacturers to invest in reducing CO_2 emissions and pollution. As a result, a disruptive technology step is required that will enable the manufactures of cars and marine engines and in general for all heavy industries to meet the forthcoming legislative standards. One very attractive way of achieving this objective is to generate power from the internal combustion engine (ICE) waste heat in vehicles. By doing this, the exhaust system created will offer greatly improved environmental performance due to enhancement of fuel efficiency and reduce the emissions at a cost that is affordable to the end user.

Secondly, and due to the fact that this technology is developed around automation and smart control of industrial processes in housing, the application concerning the building of autonomous power unit sources to aliment sensors in a wide range of temperature also needs to be achieved.

According to the published results, the silicides are materials with initial good thermoelectric properties. Consequently, it is reasonably permissible that these materials can get better properties by some changes in properties in a global

© The Author(s) 2017
J.-C. Tedenac, *Multicomponent Silicides for Thermoelectric Materials*,
SpringerBriefs in Materials, DOI 10.1007/978-3-319-58268-9_2

approach of the microstructures and structures and the research on the stabilities of compounds. These compounds present various conduction mechanisms and complex band structures. Moreover, as some of them are isotropic and some anisotropic, they could be successfully used in anisotropic thermoelectric devices. Many researches on thermoelectric properties of silicides have been fulfilled in the last 30 years, but one should remember that silicide research was initiated a long time ago by Academician Abram Fedorovich Ioffé in the former USSR [4].

This paper is divided into different chapters related to two themes: solid-state physics and chemistry of materials.

In a first section, we present a review of materials in which the physical properties are evidenced. Two kinds of materials have been widely studied: higher manganese silicides and magnesium silicides. Higher manganese silicides (HMS) MnSix exhibit interesting figures of merit at intermediate temperatures (573–873 K). Higher manganese silicides (HMS) with formula MnSix (x around 1.75) exhibit interesting figure of merit at intermediate temperatures (573–873 K) [7, 8]. These properties could also be improved by nanostructuration or doping with germanium [9, 10]. In such applications, the knowledge of the thermodynamic phase diagram is essential. This material is a p-type semiconductor. It should be associated with magnesium silicide material which is n-type. Then we describe some properties of magnesium silicide (Mg2Si) which can be doped with heavy elements and possess potentially high performances. Compared to conventional thermoelectric materials, MnSix and Mg2Si-based alloys have merit as they are non-toxic, sustainable, lightweight and low cost. We describe also some other silicides which were not deeply studied up to now, but we think that they show some interesting properties and they should be studied in the future.

The next section concerns the main originality of this paper. It is dedicated to the thermodynamic studies of multicomponent silicides. This section is divided into three subsections. Firstly, as the modern approach of phase diagrams is made presently by a global approach of phase diagrams, phase equilibria and phase stabilities, the philosophy of this approach is described. It is given by the study of thermodynamics of materials. Phase diagrams are well known as a description of a pressure, temperature and phase quantities (p, T, Niph) of the phase relationships in a system, but usually they are described as pictures coming from assessed experimental results. Sometimes these descriptions are scarce or wrong. Nowadays, the thermodynamic properties must be described in a modern way. Consequently, one should add the necessary information in the system modelling. Then, the binary, ternary and multicomponent systems containing thermoelectric materials are described by using those modern tools associated in the general CALPHAD approach which include thermodynamic calculations. The phase diagrams which will be described are those presented in the section concerning properties. They are classified into two subsections.

Finally, this monograph describes some peculiarities of silicide-based materials for thermoelectric applications. These materials are promising for many applications; some of them show very high thermoelectric figure of merit, even higher than other compounds.

Chapter 3
Review of Materials

Silicides as thermoelectrics were proposed a long time ago by E.N. Nikitin in a paper published in 1958 [5]. Since this period and due to the large use of bismuth telluride alloys, the researches were slowly growing up to 2000 years. In this last period, the scientific community has shared the societal problems concerning energy saving and environmental problems concerning the harmfulness and material recycling. Then a wide panel of materials were explored in order to replace tellurides and selenides. The main characteristics of these materials are due to two things: a high density of states (d ~ 10^{21} cm^{-3}) and a low carrier mobilities (μ ~ 10 cm^2 V^{-1} s^{-1}). In order to enhance the thermoelectric properties, it is necessary to study the decrease of the density of states and to obtain higher mobilities. It can be obtained by working on the chemical compositions and on the micro(nano)-structuration on the samples. Nevertheless, thermoelectric silicides possess high melting point and different types of conduction, and due to their heat of formation ($\Delta_f H$), they show a good thermal stability and a maximum thermoelectric figure of merit (ZTmax) [6]. Moreover, and according to the A.F. Ioffé theory [4], most of the materials are narrow bandgap semiconductors; the energy gap (Eg) presents suitable values for thermoelectric applications.

Presently the highest achieved ZTmax value for higher manganese silicide (HMS-MnSi$_{1.7}$) as p-type materials and for Mg$_2$(Si, Sn) alloys as n-type was demonstrated in several recent programmes around the world (including industrial research) to be over 1. These materials can be competitors with bismuth telluride alloys particularly in the temperature range 300–500 °C. According to the fact that they are made with abundant, cheap and recyclable elements, their interest is growing up. The only problem which should be solved concerns the module technology (electric contacts, diffusion barriers, corrosion resistance, etc.), but with time it will be done at the industrial scale.

The physico-chemical parameters (crystal structure parameters and heat of formation) of these materials are presented in Table 3.1. In this table, one can see that only CoSi and Mg$_2$Si have cubic symmetry, all other thermoelectrics with higher

© The Author(s) 2017
J.-C. Tedenac, *Multicomponent Silicides for Thermoelectric Materials*,
SpringerBriefs in Materials, DOI 10.1007/978-3-319-58268-9_3

Table 3.1 Physical characteristics of the materials

Material	Melting temp.	$\Delta_f H$ (kJ/MOL of at)	Eg	Crystal structure	Refs.
$CrSi_2$	1763	108	0.7	Hex.	
				a = 0.4431 c = 0.6364 a = 0.4442 c = 0.6366	[7, 10]
$MnSi_{1.7}$	1430	33	0.66	Tet.	
				a = 0.5525 c = 1.7463 a = 0.5526 c = 1.7517 a = 0.518 c = 4.8136 a = 0.531 c = 6.5311 a = 0.5115 c = 11.36	[6]
$FeSi_2$	1490	74	0.87	Ortho.	
				a = 1.1074 b = 0.8957 c = 0.5533	[24]
Ru_2Si_3	1970	134	1.1	Ortho.	
				a = 1.1074 b = 0.8957 c = 0.5533	[24, 75, 76]
$^{ReSi}1.75$	2213	70	0.16	Tric.	
				a = 0.3138 b = 0.3120 c = 0.7670 alpha = 89.9	[75]
CoSi	1700	100	16	Cub.	
				a = 0.44445	[39]
$Mg_2(Si, Sn)$	1270	–	0.77	Cub.	[77]
				0.6356 6.34–6.39	[52]
				6.34–6.39	[6]

silicon content present lower symmetries, and consequently they show some anisotropy of properties [7]. Although they have different crystal structures, all these structures could be considered as various deformations of tetragonal structure. In the following section, all the properties will be summarized.

Chapter 4
Properties of Compounds

4.1 Chromium Disilicide

Very few researches are related to the chromium disilicide materials ($CrSi_2$). It has a relatively high melting temperature. Now the measured figure of merit ($ZT = 0.25$) is low but achieved in undoped material and it can be supposed that enhancement of ZT can be done by doping and/or nanostructuration. The addition of Al as dopant was firstly done; it entails an increase of ZT. The thermoelectric properties of undoped $CrSi_2$ material are shown in Fig. 4.1 according to [7] and these results were confirmed later in [8]. To our knowledge, there are no results on the systematic study of thermoelectric properties of doped chromium disilicide; only few references show significant changes in the behaviour of the material. Some papers have mixed doping effects and nanostructuration, but the power factor values do not change very much [9, 10]. One more problem for $CrSi_2$ is that it is hexagonal, and due to this crystal structure, it shows significant anisotropy of Seebeck coefficient. One has a similar problem as in bismuth telluride materials [11].

4.2 Higher Manganese Silicide

The main problem existing for the so-called higher manganese silicides (HMS) concerns the phase definition. This material contains a number of crystal structures which were determined in a narrow homogeneity region ($MnSi_{1.7-1.75}$) [12–14]. The structures are tetragonal and are described with a general formula $MnSi2_{n-m}$. They are related to the Nowotny chimney ladder phases (NCLs) [15], and they differ from the point of view of crystal structures. The properties of these materials can be approached by using the VEC (valence electron counts per number of transition element) theory which is described in reference [16]. The valence electron count (VEC) corresponds to the number of valence electron per formula unit taking part

© The Author(s) 2017
J.-C. Tedenac, *Multicomponent Silicides for Thermoelectric Materials*,
SpringerBriefs in Materials, DOI 10.1007/978-3-319-58268-9_4

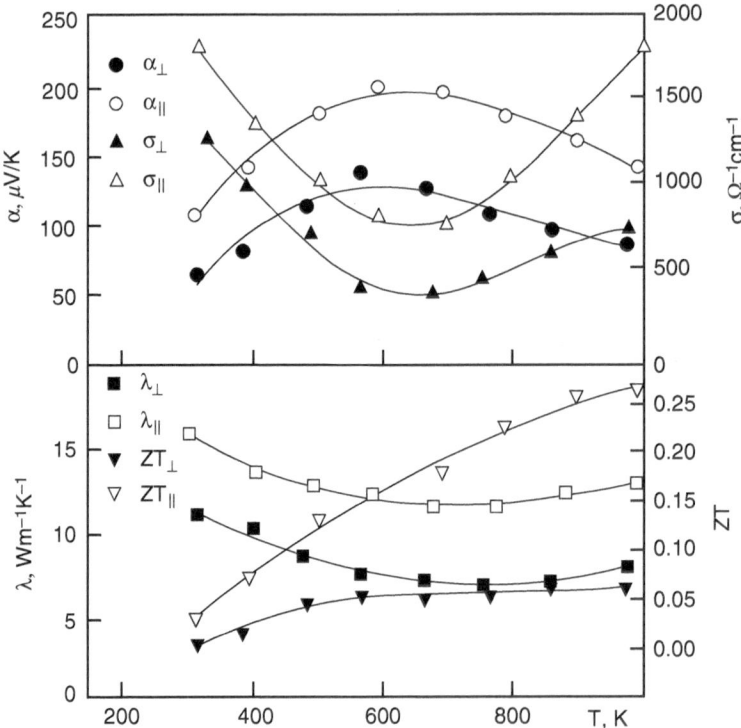

Fig. 4.1 Thermoelectric properties as a function of temperature of an undoped CrSi$_2$ crystal taken from the Ref. [8]. This figure shows the behaviour difference in each direction of a crystal. This difference is growing in the temperature range (600–900 K)

to the chemical bonds. For example, for composition MnSi$_{1.7361}$, the number of electrons participating to the valence band is VEC = 13.944. In such a count, the compound having less than 14 values means that the corresponding material is p-type. When VEC is over 14, the material is n-type. The question of properties finally depends on the composition in use, and in this compound, it is easy to proceed to a band engineering by little changes in the chemical composition. Another feature is that in all published papers it has been shown that a second phase precipitates in the alloy. This was evidenced by [17] in the material of composition MnSi$_{1.2}$ during the cooling stage. Ingots obtained by direct cooling present a thin-layered microstructure [17]. Annealing is necessary to remove this problem, and after a treatment at 1353 K, the microstructure remains homogeneous. This problem occurring at crystallization looks like an antiphase boundary. Nevertheless, one positive thing is that these precipitates do not destroy the material orientation and structure and do not affect the thermoelectric properties.

These precipitates are plates oriented orthogonally to the tetragonal axis of the structure [7]. All blocks of phase are oriented in the same way, so in neutron diffraction study such a sample is observed as a single crystal.

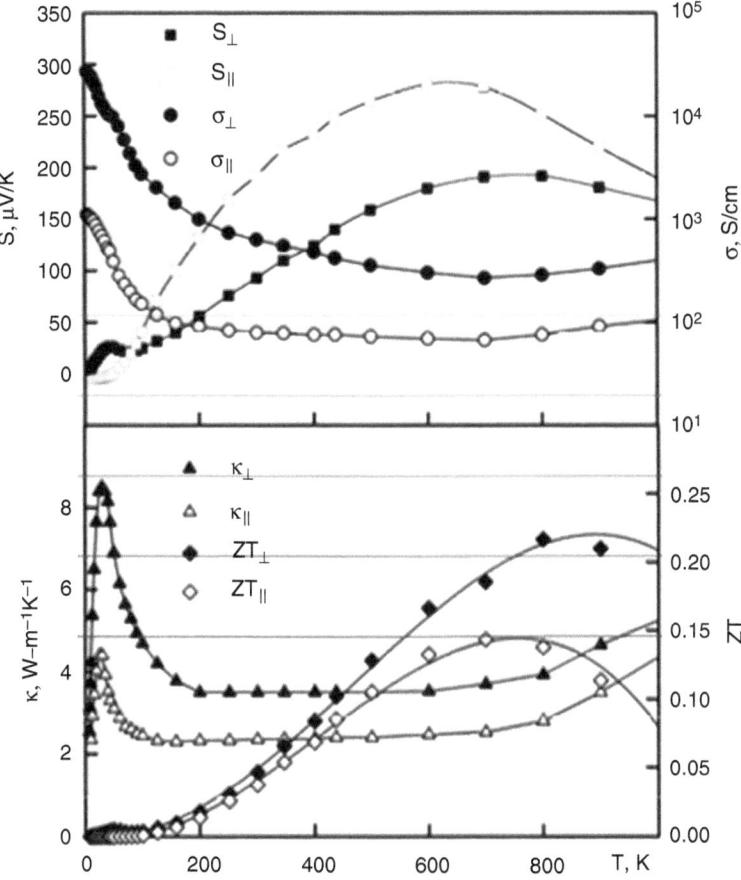

Fig. 4.2 Thermoelectric properties (electrical conductivity, Seebeck coefficient and thermal conductivity) measured in two directions in HMS crystal from [6]. As a function of the orientation, the material presents difference between the orientations (perpendicular and parallel show differences becoming significant at high temperature)

Similar to $CrSi_2$, HMS has significant anisotropy of Seebeck coefficient that should be taken into account in some to use it in thermoelectric module shaping [7]. Figure 4.2 shows thermoelectric properties of pure HMS.

In literature, the thermal conductivity of HMS-based solid solutions is described by the formula $MnSi_{1.7}$ which is considered as pure compound. As a consequence of anisotropy, the thermal conductivity of all solid solutions is higher than that of pure HMS along the c-axis and lower than that across c-axis. Because of that, only the change of carrier concentration allows to increase thermoelectric figure of merit. Depending on the orientation, the precipitation of the second phase can be considered as a system of energy barriers for current and heat flow along c-axis. It was reported in [18] that, by using a simple model, the so-called mobility edge, one can suggest that for the energy values as E < Ec, the mobility of current carriers is negligibly low, and at some energy E > Ec, the mobility is equal to usual band mobility. The transport properties of such materials were studied and it goes to a high value of the power factor, and it is explained that the electronic and thermal conductivity is higher when the Fermi level is closed to conduction band edge (Ec) [18]. This is typically done in semiconductors by controlling the composition of alloys and described in the so-called bandgap engineering which is the process of controlling or altering the bandgap of a material. This system of energy barriers can be changed by the thermal and/or engineering and doping [19]. Optimization of the composition of HMS allowed to improve its thermoelectric properties.

4.3   Iron Disilicide

Iron disilicide is one of the oldest studied thermoelectric materials. It is a cheap material, but the problem of phase transformation at high temperature leads to some difficulties in the material synthesis. Since Ware and McNeil suggested to use it in thermoelectric devices [20], a lot of researchers wrote many papers on this material essentially by Birkholz et al. as in [21]. At high temperature iron silicide (stable at T > 1210 K) is named α-phase, and it is metallic with tetragonal structure. This α-phase corresponds to a stoichiometric iron disilicide ($FeSi_2$). At low temperature, it is named β-phase; it is semiconductor in an orthorhombic structure. The change of crystal structure is done with a slight difference in stoichiometry. The low temperature is nonstoichiometric: $LT\text{-}FeSi_{-2.4}$. The temperature of 1210 K corresponds to a decomposition reaction.

The main feature of the majority of Seebeck coefficient measurements is a sharp increase of absolute value of Seebeck coefficient at 100 < T < 300 K (Fig. 4.6). This corresponds to the temperature of optical phonon excitation, so it is allowed to suggest that such a growth results to the drag of current carriers by optical phonons. This statement is not directly confirmed, but the work of [22] shows a good agreement with experimental data [22, 23].

This material has been used in the past as thermal sensors for high temperature. This application requires the availability of active material in terms of high sensitivity, linearity and functional long-term stability, high-temperature stability of junctions between active sensor material and metallic connectors as well as thermomechanical stability between carrier structure, semiconductor and coatings. Suitable calibration and test apparatus are required for such system evaluation.

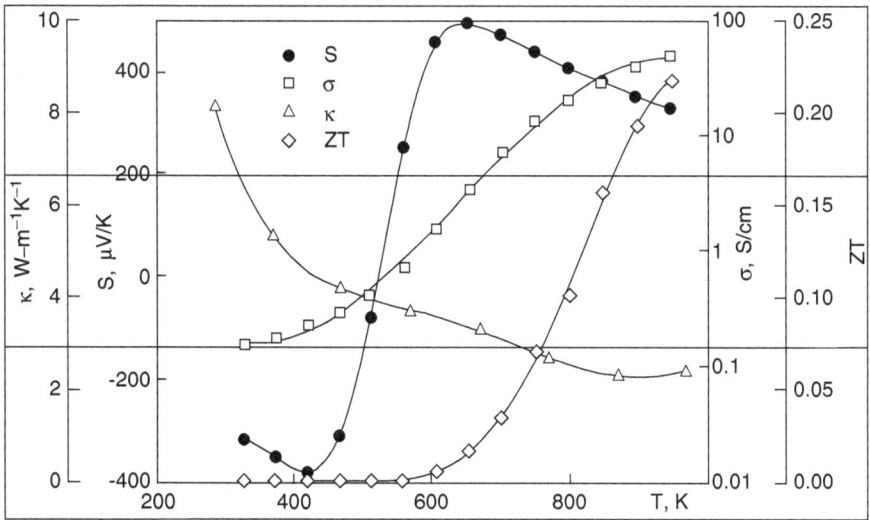

Fig. 4.3 Thermoelectric characteristics of the ruthenium sesqui-silicide (Ru$_2$Si$_3$, (Sou 2002)). This material changes the conduction type at 500K and presents the highest ZT values above 800K

4.4   Ruthenium Silicide (Ru$_2$Si$_3$)

The ruthenium sesqui-silicide Ru$_2$Si$_3$ is very similar to iron disilicide [7]. It has a phase transition at ~1240 K. The low-temperature phase α-Ru$_2$Si$_3$ possesses an orthorhombic structure, whereas the high-temperature phase β-Ru$_2$Si$_3$ has a tetragonal one. Both phases are semiconductors, and we will see that the phase diagram is not very simple around the melting point of the compound as it was described in [24]. Similar to HMS materials, the crystals of α-Ru$_2$Si$_3$ show platelike precipitates inside the matrix. In the publications [25, 26], the precipitates are orthogonally directed to the direction (010) axis. But some other possibility can appear, depending on the crystal growth procedure [26, 27]. Concerning the stoichiometric Ru$_2$Si$_3$ compound, the thermoelectric properties were measured by using a single crystal having with different growth rate and possessing domain structure. All domains had the same orientation of (010) axis with some (100) and (001) intermixed orientations. Figure 4.3 shows the thermoelectric properties of these crystals [36]. High enough values of figure of merit for doped Ru2Si3 have been presented in a number of papers [24, 27, 28, 29, 30]. Practically all of them have elements of extrapolation for ZT calculation, because the data obtained from different samples were used. Nevertheless, systematic study of this material has not fulfilled yet.

This compound has been used by association with Mn_4Si_7 in order to build new chimney ladder phase [31, 32]. The thermoelectric properties of the directionally solidified $Ru_{1-x}Mn_xSi_y$ alloys ($0.55 < x < 0.90$) have been investigated. The ZT value for a crystal with $x = 0.90$ was as high as 0.76 at 874 K, which is a very good value in this case, but this result needs to be reproduced and confirmed.

4.5   Rhenium Silicide ($ReSi_{1.75}$)

Semiconducting rhenium silicide $ReSi_{1.75}$ is known as a narrow bandgap semiconductor [33]. The dimensionless figure of merit, ZT, has been estimated to have a high value for the stoichiometric material; however due to some difficulties in the synthesis, this has not yet been confirmed. The prospect of achieving high values of ZT looks rather promising since the discovery of very high carrier mobility of 370 cm^2/Vms at room temperature in $ReSi_{1.75}$ [35], but this mobility value has not been confirmed later.

During many years, this compound was considered as stoichiometric ($ReSi_2$) and has a tetragonal structure (space group 4/mmm, a = 0.3131; c = 0.7676 nm [42]). Recently it was shown that defects are present in huge quantity and causes nonstoichiometry and a symmetry lowering for the crystal. Then the structure is really less symmetric. The crystal structure of $ReSi_{1.75}$ has been analysed as triclinic (space group P1, a = 0.3138; b = 0.3120; c = 0.7670 nm; $\alpha = 89.90$ by [35]). This explains that the thermoelectric properties of $ReSi_{1.75}$ are highly anisotropic [36]. The electrical conductivity is of n-type when measured along (001), while it is of p-type when measured along (100). The value of the Seebeck coefficient along (100) is moderately high (150–200 μV/K) and grows up to 250–300 μV/K along the (001). As a result, a very high value of dimensionless figure of merit (ZT) of 0.7 can be achieved at 1073 K by using appropriate orientation [33]. Similar results obtained later are shown in reference [37], where the ZT value is even increased up to 0.8 with a small amount of Mo addition (2% substitution for Re).

4.6   Cobalt Monosilicide

Cobalt monosilicide is a semimetal with a cubic structure, having high power factor $S^2\sigma$ ($\approx 50 \cdot 10^{-4}$ W $\cdot$ m^{-1} K^{-2}). The current carrier concentration can be easily changed by Co substitution with Ni or Fe. Fe addition results in the change of conductivity type from n to p at ~5% of Fe. High enough power factor is explained by multi-valley structure of CoSi conduction band. This conclusion was made after the analysis of optical reflection measurements near the plasma minimum [37], and it was confirmed by the band structure calculation [38]. Thermoelectric properties of CoSi are shown in Fig. 4.4 taken from [39]. It seems that the method of synthesis is very important for the change of ZT in this case. As the melting temperature is relatively high, the more suitable synthesis method is the powder metallurgy [40].

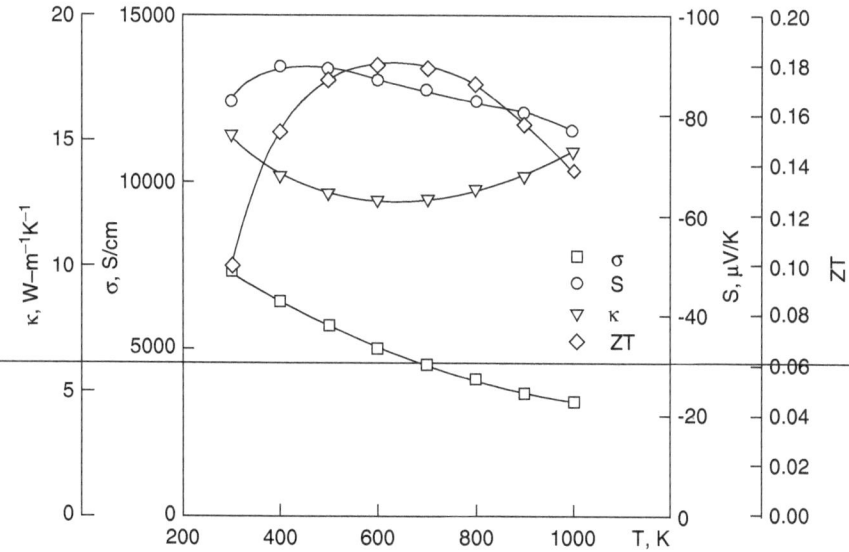

Fig. 4.4 Thermoelectric properties of pure CoSi according to the Ref. [39]

4.7   Magnesium Silicide and Related Compounds

It was shown in the 1960s that the Mg_2X compounds (X = Si, Ge, Sn) and their solid solutions should be promising compounds for thermoelectric energy conversion [41–43]. Since that time, in the last century, very few researches were made on these materials, even if these previous results showed interesting electronic properties.

Since 15 years, a keen interest for this compound is sustained around the world. Nowadays, the interest in these materials is fuelled by the very high values of thermoelectric figure of merit obtained in some of these researches and by the concept of sustainable material applied to this compound. It is made with abundant raw constituents, and it has low density, allowing manufacturing of light, sustainable and relatively cheap devices. Some years ago, really good thermoelectric material has been produced in the Laboratory for Physics of Thermoelements in Ioffe Physical-Technical Institute on the base of solid solutions of these compounds [44].

The great interest of these materials rests on the crystalline structure which is a very symmetrical one and which allows numerous substitutions on the different crystallographic sites, leading to a number of multicomponent alloys (interesting for band engineering). This made possible to obtain such high ZT values (1.2–1.3), and it is what we observed in the literature. By alloying magnesium silicide with germanides and/or stannides, a lot of interesting results have been produced around the world. A second interest in Mg_2X compounds is the complex structure of conduction band (Fig. 4.5, [45]). Minimum of conduction band is located in X-point of Brillouin zone, and, hence, conduction band is multi-valley band that is favourable

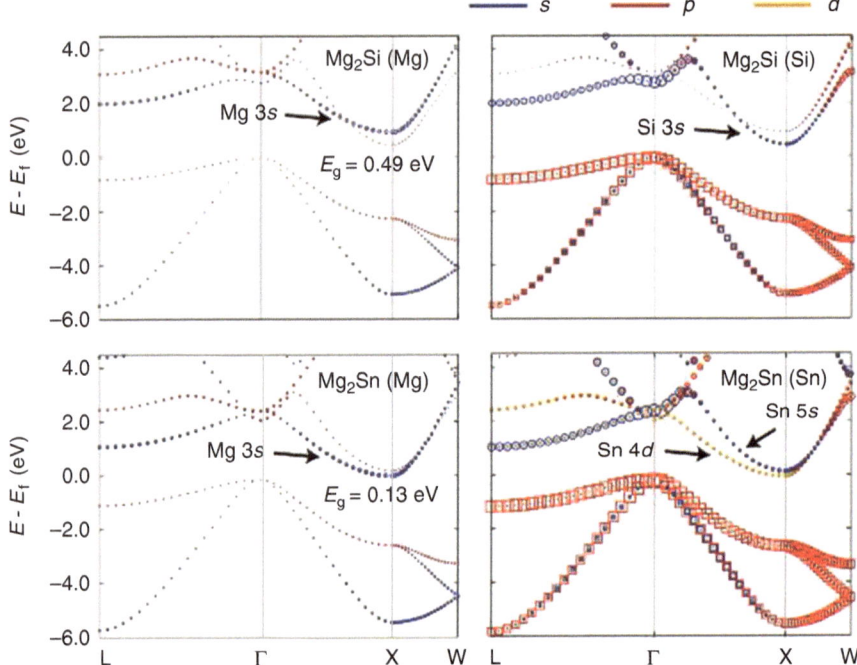

Fig. 4.5 The band structure of the Mg$_2$X (Si, Sn) compounds from the Ref. [45]

for thermoelectrics. Two subbands of the conduction band are separated by small gap ($\Delta E < 0.5$ eV), and they have been created from states of Mg and of the elements of the IV-group. The lowest subband is created by Mg states in Mg$_2$Sn, but in Mg$_2$Si and Mg$_2$Ge, it is created by Si(Ge) states. In the solid solutions of Mg$_2$Si or Mg$_2$Ge with Mg$_2$Sn, there is a subband inversion.

In Mg$_2$Si$_{0.4}$Sn$_{0.6}$ solid solution, the gap between subbands practically is equal to zero. In this case the electrons of both subbands take part in transport phenomena that result in thermoelectric figure of merit increase [46]. Another advantage of this material is the absence of interband scattering in these materials [47].

A lot of researches were developed in order to enhance the ZT values: by alloying, nanostructuration, precipitations, etc. The optimization of solid solution concentration and electron concentration allowed to develop very efficient thermoelectric properties [50–57].

Figure 4.6 shows thermoelectric properties of n-Mg$_2$Si$_{0.4}$Sn$_{0.6}$ solid solution with optimum electron concentration. Such a solid solution has average figure of merit in the temperature range 320–850 K of ZT = 1.2. It allows to use it in single-stage thermoelectric devices working in the mentioned temperature range. Much higher coefficient of performance in this temperature range has been obtained (up to 1.2–1.3).

SiC nano-powders and nano-wires were introduced in Mg$_2$Si-based material such as Mg$_{2.16}$(Si$_{0.3}$Sn$_{0.7}$)$_{0.98}$Sb$_{0.02}$. They have excellent toughness as well as high

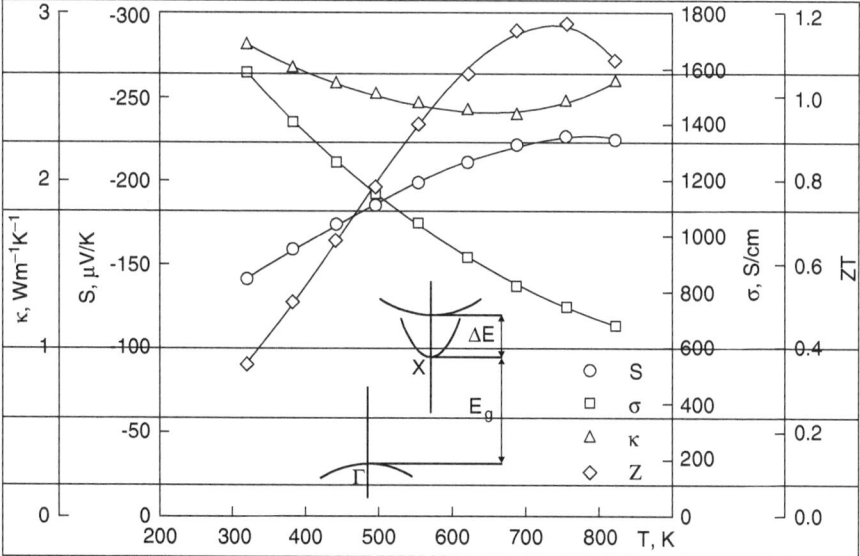

Fig. 4.6 Thermoelectric properties of the solid solution phase n-Mg$_2$Si$_{20.4}$Sn$_{20.6}$ with optimum electron concentration. In the insertion a scheme of band structure of Mg2X compounds shows the bandgap crossing

strength. Transport properties remained inert to the presence of the SiC nano-phase and finally excellent thermoelectric performance (ZTmax value of 1.20 at 750 K was maintained) [48].

Concerning the technology of thermoelectric generators, Komatsu Ltd. in Japan was the first company to build thermoelectric modules with n-leg of Mg$_2$(SiSn) and p-leg of HMS produced [49], and since that time a huge quantity of applied and industrial programmes have been developed successfully around the world and give some interesting devices for automotive applications.

4.8   Silicides Among the Other Thermoelectrics

In order to show the place of silicide among other materials used in thermoelectrics, Fig. 4.7 shows the temperature dependencies of the figure of merit of two of the best silicide thermoelectrics (HMS as p-type and Mg$_2$Si$_{0.4}$Sn$_{0.6}$ as n-type) among the similar dependencies for the best thermoelectrics working in various temperature ranges. The base of these figures was taken from the reference [58]. As one can see from Fig. 4.6, the alloyed magnesium silicide-based material shows the best thermoelectric properties in the temperature range 600–850 K in comparison with the other thermoelectrics of n-type. The situation with silicide thermoelectrics of p-type is not so good, but future researches will allow to improve it.

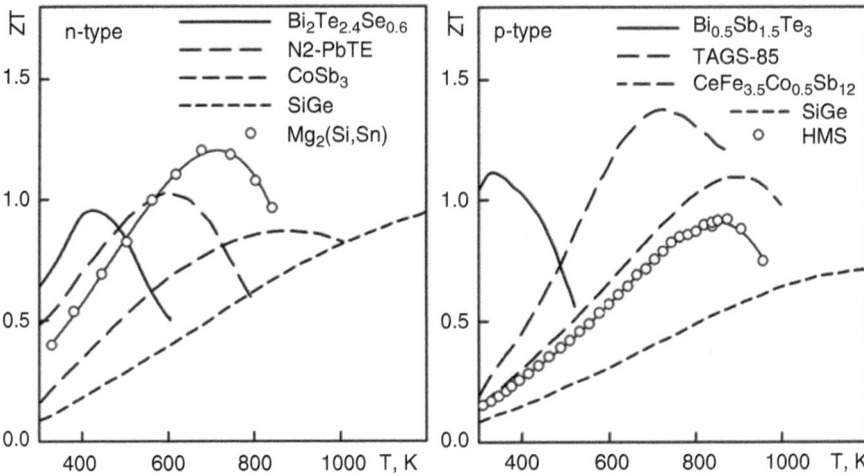

Fig. 4.7 The thermoelectric figure of merit of the representative silicide thermoelectrics in comparison with the other materials, having the highest ZT in their temperature range according to [58]

Chapter 5
The Thermodynamic Approach in Thermoelectric Materials

5.1 Basis: The Modern Approach of Phase Diagrams

The main interest of thermodynamics of materials is made presently by a global approach of phase diagrams, phase equilibria and phase stabilities [59]. Phase diagram is well known as a description in pressure, temperature and phase quantities (p, T, N_{iph}) of the phase relationships in a system. Phase equilibria describe the phases in a system which are in equilibria in certain p, T ranges through the Gibbs-Duhem equation. Phase stabilities are described by the free energy function of phases in a system at 0 K and in the whole range of temperature, and it is depending on the lattice stabilities [59]. Thermodynamics of materials is nowadays mainly described by the CALPHAD method which is globally explained in four excellent books [3, 59, 60, 61]. In the last years, this method has been widely developed for metallic systems, oxides and semiconductors, for example, gallium arsenide, cadmium telluride and lead telluride [62–65].

Basically the calculation of phase equilibria in a multicomponent system is obtained by the minimization of the total Gibbs energy, where G is a summation of the Gibbs energy of all phases taking part in each equilibrium as is expressed by Eq. (5.1):

$$G = \sum_{i=1}^{p} n_i G_i^0 = \min.$$
(5.1)

The thermodynamic description of the whole system requires the assignment of thermodynamic functions to each phase. The main interest of the CALPHAD method is the use of a variety of models to describe the free energy functions of the different phases taking part to equilibria. The Gibbs energy function of a phase i can be written as in Eq. (5.2):

© The Author(s) 2017
J.-C. Tedenac, *Multicomponent Silicides for Thermoelectric Materials*, SpringerBriefs in Materials, DOI 10.1007/978-3-319-58268-9_5

$$G_m = \sum_i x_i \left({}^iG_0 - {}^iG_0^{\text{phys.}} \right) + G_m^{\text{phys.}} - T\, S_m^{\text{ideal.}} + {}^EG_m \tag{5.2}$$

The effects of particular physical phenomena such as magnetism are taken into account by subtracting them from the description of the endmembers (reference state), and it is introduced for the solution through the contribution G_m^{phys}. The last term on the right-hand side is the excess term, representing the contribution of physical and chemical interactions in the system.

Usually the temperature dependence of the Gibbs energy is expressed as power series of T where a, b, c and dn are coefficients and n are integers (Eq. (5.3)). The pure elements are represented by the same function, and they are tabulated in the database of A. Dinsdale [65]:

$$G = a + b.T + c.T.\ln(T) + \sum dn.T^n \tag{5.3}$$

By taking into account all the parameters, the Gibbs energy of phases is then represented by three contributions in which all previously described contributions are included and expressed in the Eq. (5.4):

$$G^0 = {}^{\text{ref}}G^0 + G^{\text{ideal}} + G^{\text{xs}} \tag{5.4}$$

The first term of the right-hand side corresponds to the Gibbs energy of the components, the second one corresponds to the entropy of mixing for an ideal solution and the third term, the so-called excess term, represents all the deviations from ideality.

Thermodynamic modelling of phases is the core of the CALPHAD approach. Most systems have few strictly stoichiometric compounds. Some phases are deviating from the ideal stoichiometry, and it is necessary to take this problem into account. The most used models for solution phases are random substitutional and ordered sublattices. In ordered solid phases, the Wagner-Schottky model is mainly used for describing small deviations from stoichiometry (which are noninteracting defects). Many intermetallic binary compounds are generally nonstoichiometric and can exist in a large range of composition. In this case the most used model is the sublattice model [3]. In the particular case of thermoelectric materials, as in semiconductors [8], it is important to model the phases according to the crystal structure and including the structure defects. Moreover, additional information from physical properties (chemical potentials of electrons and holes measured by transport experiment) can be put inside this model and implement the database.

All the models can be used for ternary and multicomponent systems. In this case, it is necessary to add high-order interaction terms in the expression of the excess free energy in order to take into account the unavoidable interactions between components, particularly in the liquid state.

In such high-ordered systems, the Gibbs energy must be calculated from extrapolation of the excess quantities of constituent subsystem as shown in Fig. 5.1.

$$G = \sum x_i G_i^0 + RT \; x_i \ln x_i + G^{ex}$$

BINARY	*Assessment* G_{bin}^{ex}
	⇓
TERNARY	*Extrapolation* G_{bin}^{ex} + *Assessment* G_{ter}^{ex}
	⇓
QUATERNARY	*Extrapolation* $(G_{bin}^{ex} + G_{ter}^{ex})$ + *Assessment* G_{qua}^{ex}
	⇓
HIGH ORDER	*Etc ...*

Fig. 5.1 Description of the CALPHAD procedure for the calculation of high-ordered systems [62]

In fact the relationship between chemical potentials of the phases and the system gives nonlinear equations that can be solved by numerical methods such as Newton-Raphson methods [60]. The CALPHAD-type packages use mathematical methods to minimize the Gibbs functions. It uses different data sources for the assessment of the systems going from classical thermal analysis to ab initio calculations.

Experimental data used for these calculations are those concerning the phase diagram measurements, enthalpy of formation of the compounds, enthalpy of mixing and EMF measurements, thermal analysis and X-ray diffraction. The results of ab initio calculations are also useful for using as input in the calculations, and they are taken into account with time. Those results are done at 0 K, but they can be extrapolated at 298 K, according to the fact that the temperature dependence in this temperature region is considered as negligible. The Gibbs energy of pure elements is taken with reference to the enthalpy of the elements in the SER state (standard element reference) and the elements in their stable state at 10^5 Pa and 298.15 K. This description is adopted from the SGTE database [65].

The purpose of this paper is not to describe in detail the method but to underline the applications. For a more precise knowledge of the problem, the reader can refer to the books which were presented in the first part of this chapter.

5.2 The Binary Systems

The main systems containing thermoelectric materials have been studied many times ago, and also recently due to several interesting properties (tellurium free, higher melting temperature and thermal stabilities), any thermodynamic properties were pointed out.

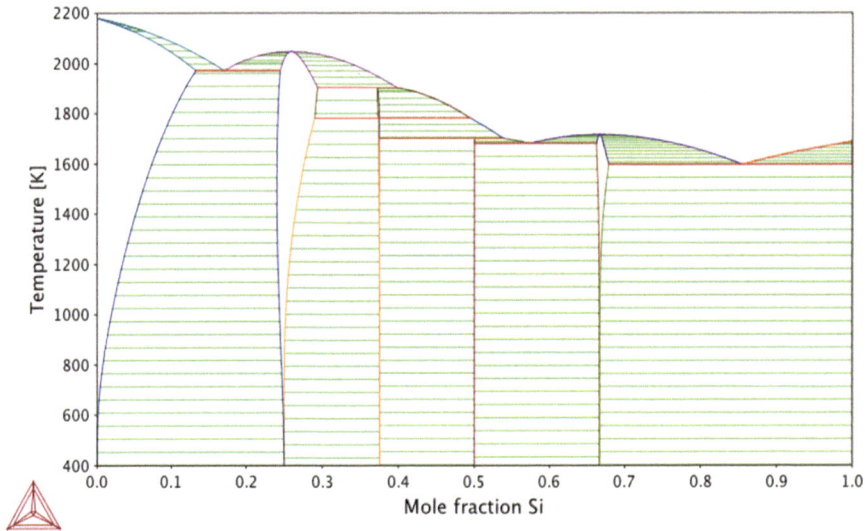

Fig. 5.2 The phase diagram of chromium-silicon. The compound CrSi₂ is characterized by a congruent melting point and a departure to stoichiometry at high temperature

The materials in question have been described in Sect. 5.1. By taking as an example one of these compounds chromium silicide (CrSi₂), one observed that the ZT values in the temperature range 700–800 K are around 0.2–0.3, while the initial value of the raw compound is around 0.1. These differences are due to the composition change and microstructure change (as precipitation of second phase). A second example concerns the manganese silicides. In this case the situation is a little more complex as the initial composition is ranging around 1.7–1.8% of Si. It is why the formula MnSi$_{x\,is}$. The change of thermoelectric properties in these materials must be done by using multicomponent alloys. These two examples show that this enhancement of properties is due to a change of the chemical compositions intentionally done and a thermodynamic study of materials is needed.

Concerning the Mg₂(Si, Sn) materials, it is very difficult to resume all the researches made on these alloys. Generally the ZT factor maximum is obtained for doping with Sb (sometimes with Bi), and multicomponent alloys are constructed on the base of a reciprocal ternary Mg₂Si-Mg₂Sn-Mg₂Ge. A value of ZT over 1.2 has been obtained many times by several teams around the world as in [50, 51].

The applications of these systems show that final materials must be made with more than three elements. In some systems the analysis of the phase stabilities can be understood by using a binary phase diagram, but in most cases, the problems to be solved need the whole study of the ternary, quaternary and more component systems.

In Figs. 5.2, 5.3, 5.4, 5.5, 5.6 and 5.7, we present the phase diagrams of four systems which can be considered as significant for the study of thermoelectric materials. This is not an exhaustive list because the purpose of this paper is not to

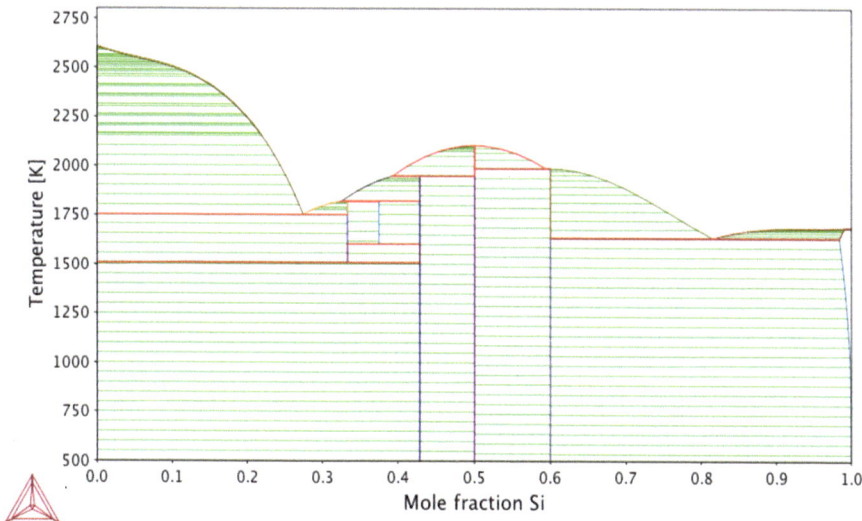

Fig. 5.3 The phase diagram of ruthenium-silicon. The compound Ru_2Si_3 is characterized by a congruent melting point close to the eutectic reaction

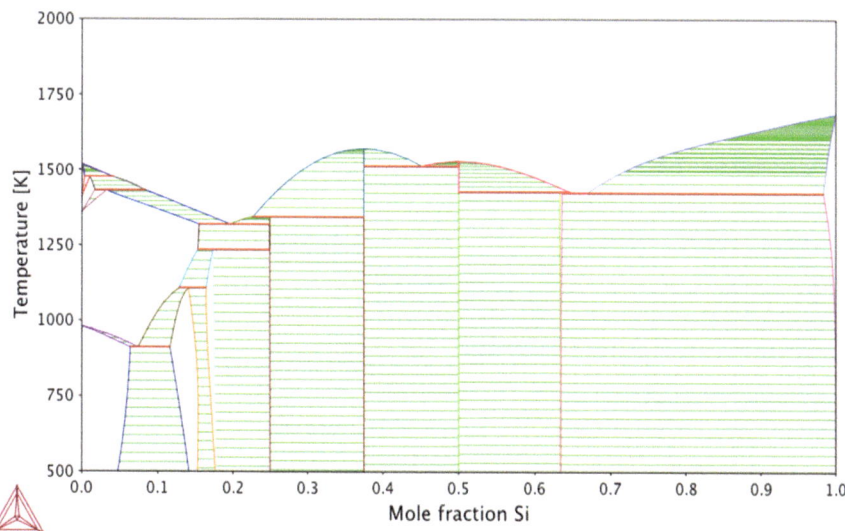

Fig. 5.4 The phase diagram of manganese-silicon

show all of those which are interesting for thermoelectricity, but rather some representative ones where we can point out the particularities (Ru-Si, Cr-Si, Mn-Si, Mg-Si). They have been chosen as they summarize all the particular phenomena which are influencing the synthesis and behaviour of these materials.

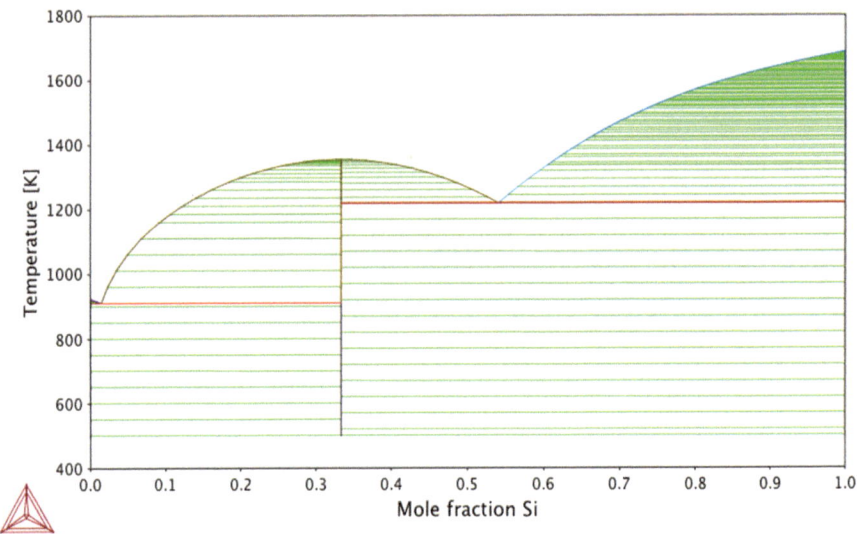

Fig. 5.5 The phase diagram of magnesium-silicon

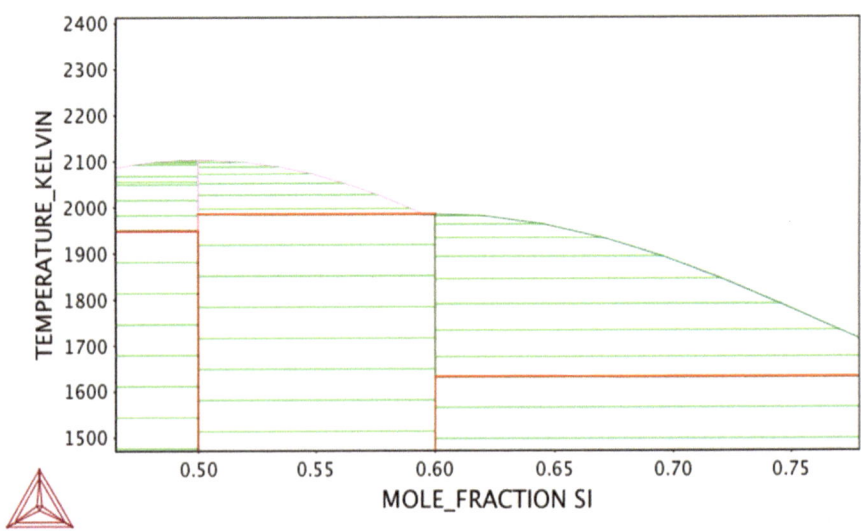

Fig. 5.6 Partial Ru-Si phase diagram. This figure presents the melting and/or crystallization conditions for Ru_2Si_3. The possibility to crystalize a single phase is impossible to do because melting temperature of compound and eutectic are collapsing

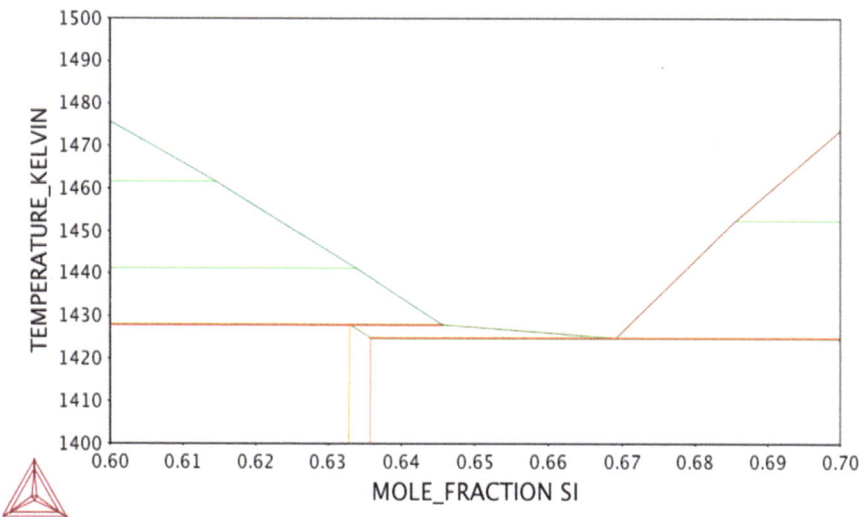

Fig. 5.7 Partial Mn-Si phase diagram. This figure presents the melting and crystallization conditions for HMS phase (MnSi$_x$). The crystallization in a single phase is impossible to do by solidification. A too small temperature interval exists between MnSi$_x$ decomposition and eutectic

The two compounds CrSi$_2$ and Mg$_2$Si possess congruent melting points. A contrario MnSi$_x$ has a peritectic decomposition temperature (Fig. 5.7). Concerning Ru$_3$Si$_2$, it is in a very strange situation where the melting point is at the same value as the closest eutectic. The first step of system's understanding is the reality of the phase compositions. In such intermetallic compounds, two kinds of problems occur. The first is related to thermal defects which are activated at high temperature, and their number is bigger as the temperature is high; this is the case for Cr and Mn systems. By taking into account this fact, the phase diagrams show a certain non-stoichiometry at high temperature. Concerning the materials with very high melting temperature (Ru or Re), this problem is slightly different, the zone where it will happen is very far from the temperatures where they are used in thermoelectricity, and the compounds can be considered as stoichiometric. Concerning MnSi$_x$ one observes a departure to stoichiometry at low temperature which is coming from the possibility of exchanging atoms in their crystallographic positions [52]; this particularity is observed in numerous intermetallic phases. The study of multicomponent systems is necessary to change the properties of thermoelectric materials, and they are presently under study or in development in the scope of ZT enhancement.

Secondly, a large use of the well-known Scheil-Gulliver rule [3] gives some directions for the behaviour of materials during synthesis and systems technology.

The first comment one should have concern is the problems of crystallization of materials, often the first step of synthesis.

The diagram in Fig. 5.6 shows some indications on the Ru$_3$Si$_2$ melting conditions (or crystallization). Even if this compound is congruently melting and according to the former remark, it is necessary to make synthesis with a small departure to

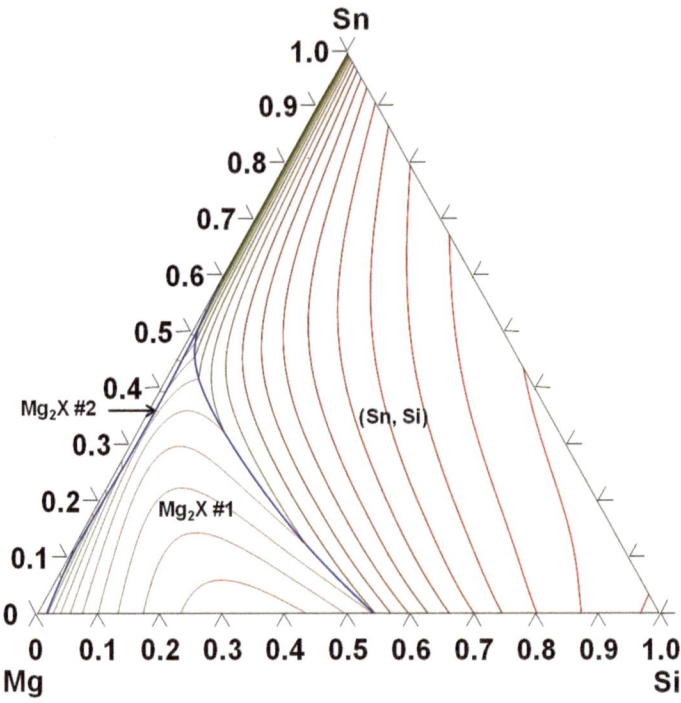

Fig. 5.8 The liquidus surface in the system Mg-Si-Sn

stoichiometry on the silicon side, in order to make sure that the samples will possess only the thermoelectric material. In this case one cannot remove the very low pre-cipitations of Si, which are not really pregnant for thermoelectricity. Concerning CrSi$_2$, the situation is simpler, the melting temperature is lower and the compound is congruently melting with a nice shape of liquidus, and then the crystallization is easiest to do, as well for the Mg$_2$Si compounds which have same characteristics but at much lower temperature. The hardest situation concerns MnSi$_x$-based materials. In this last case (Fig. 5.7), two problems are present. First is the small difference in between the peritectic decomposition temperature of compound and the eutectic temperature. Secondly, the fact exists that some polytype inside the range homoge-neity entails the plate precipitation of a second phase in the matrix. In this last case, the problem is solved with annealing treatment (see first section).

5.3   The Ternary Systems

As explained previously, ternary system descriptions are useful for understanding multicomponent materials, but for the non-specialists in thermodynamics, it is often complicated to well understand the problems. The approach of the ternary systems

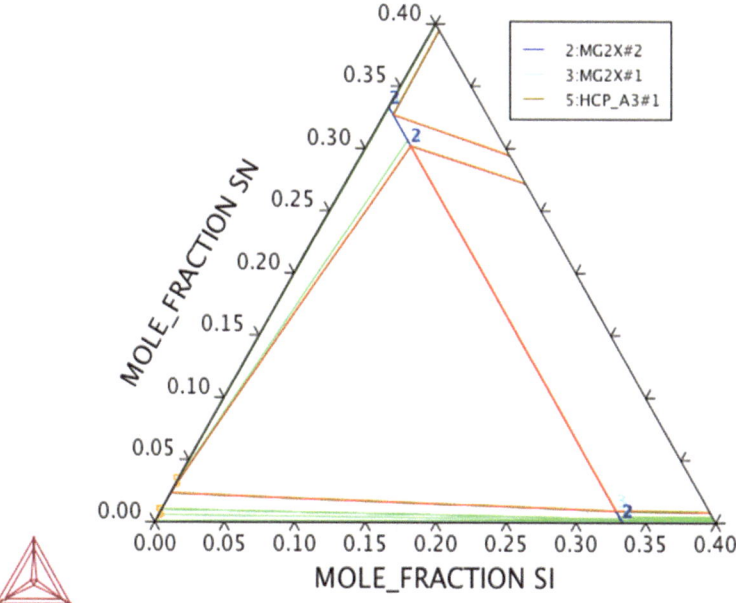

Fig. 5.9 The partial isothermal section of Mg-Si-Sn at 750 K shows the phase separation between the two solid solutions Mg₂X named Mg₂X#1 (Ge rich) and Mg₂X#2 (Si rich)

must be done step by step. It can be firstly understood by the use of liquidus projections. A liquidus projection is the representation of the surface where the first crystallizing phase is shown. According to the Gibbs phase rule, it's a description of the reaction scheme in a system. This liquidus surface is the surface limit between liquid phase and all the two-phase fields liquid + solid. As it is a representation of the first crystallizing field, finally it is a description of the phase stability of all phases in this temperature range. What is the information which can be obtained from this? It will be seen from a few examples.

As a first example, in the ternary Mg-Si-Sn, two different solid solution phases (based on Mg₂SiSn and Mg₂SnSi) are crystallizing in different ranges of temperature (Fig. 5.8). It shows a difference of phase stabilities in the ternary system (these two phases are coexisting in solid state at low temperature inside a miscibility gap; see Ref. [3]). One can observe that the silicon-based ternary solid solution is more stable at high temperature than the tin-based one (Figs. 5.8 and 5.9).

The first concerns the phase separation in the Mg2Si-Mg2Sn pseudo-binary systems studied in [53, 54]. In this system it was shown higher that two different composition phases are precipitating in the solid state. In the first figure, the partial isothermal section of Mg-Si-Sn, we present the phase stabilities in the section at at.% of Mg = 0.666. The phase MgX#2 is the Sn-rich phase (Mg{Si, Sn}) and MgX#1 is the Si-rich phase (Mg{Si, Sn}), and in between there is the mixture of them (Fig. 5.9).

The second example concerns the HMS materials. As the materials MnSix (HMS) can be doped by Ge or Cr or by both elements as well, the interesting

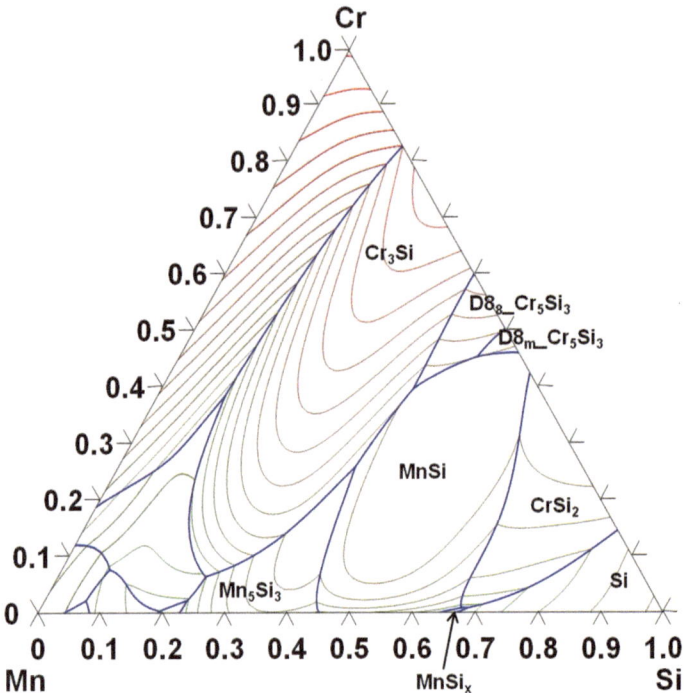

Fig. 5.10 Liquidus surface in the system Cr-Mn-Si. It shows the narrow crystallization field of MnSix in this ternary, by comparison with the crystallization fields of the other binary compounds

problem is to study the ternary systems in order to understand the behaviour of such material [56–58]. The liquidus projection is a proof of the phase stability inside the solidification range (Figs. 5.10 and 5.11). In the ternary Cr-Mn-Si, one can see that the high-temperature stabilities are related to the different fields of first crystallization. One can observe that the crystallization field of $MnSi_x$ is very small comparatively to the silicon-based phase and the MnSi compound. At this first glance, one can understand why it is so complicated to obtain a pure HMS phase.

A second kind of representation is the isothermal section. They give information on the change of phase relationships as a function of temperature. By using the isothermal sections at low temperature, this problem will be evidenced; they are shown in Figs. 5.12 and 5.13. We have chosen the temperature of 750 K (450 °C) because it is closed to the operating temperature in a thermo-generator plugged on a gas exhaust of an oil engine and of some diffusion processes slowly occurring in this region.

At this stage of discussion, we should highlight the problem of the database used for the construction of the systems. The pending problem in this thermodynamic study is to use relevant thermodynamic functions of phases and system. It is the key point of an accurate database building. Taking into account the previous researches [34–36], new results obtained in calculations and experiments, an accurate database

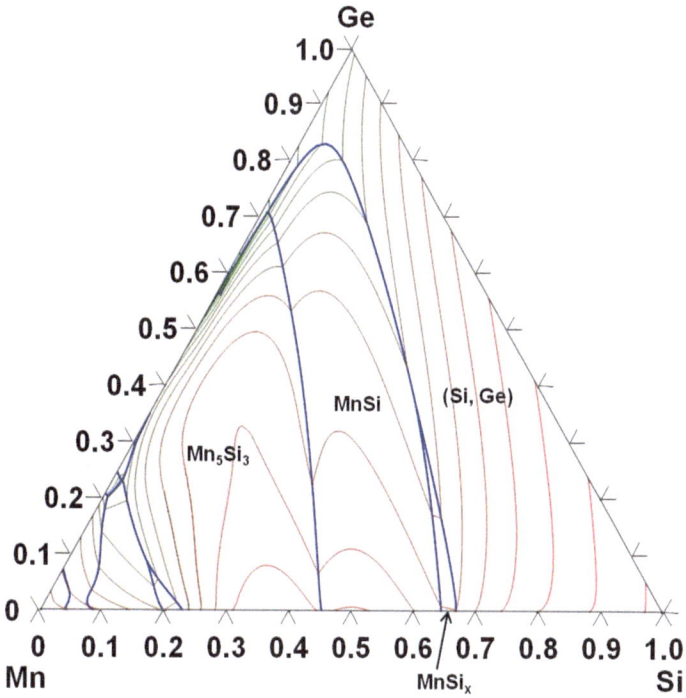

Fig. 5.11 Liquidus surface in the system Ge-Mn-Si. The crystallization field of MnSix is greater then in the chromium ternary

was constructed, serving as a guide for doping the HMS phase with both Ge and Cr atoms to improve the thermoelectric properties of this promising alloy [56–58].

Taking into account these assumptions, the database concerning Cr-Ge-Mn-Si system obey to these requirements; we can now go back to the ternary analysis. In Figs. 5.12 and 5.14, we show the whole ternary Cr-Mn-Si and Ge-Mn-Si systems where the phase relationship description and particularly around the MnSix compound is shown. This kind of figure is more useful when it is compared with a vertical section (named isopleth). An isopleth section concerning the solubility of chromium in MnSix is shown in Fig. 5.14. All equilibria from high temperature (liquid) to low temperature are shown in the section CrSi$_2$-MnSi$_x$. The comparison of the isopleth and the isothermal sections gives a good knowledge of the solubilities of Mn or Cr in the endmembers of system with respect to temperature and compositions (Fig. 5.15).

Another tool is very useful for the material synthesis; it is the Scheil-Gulliver approach which can show the crystallization scheme in multicomponent alloys (it is assumed that no diffusion exists in the solid phases, which is really the case in most of the processes in solid-state chemistry, infinitely rapid diffusion in the liquid phase which is an approximation depending on diffusion differences between elements

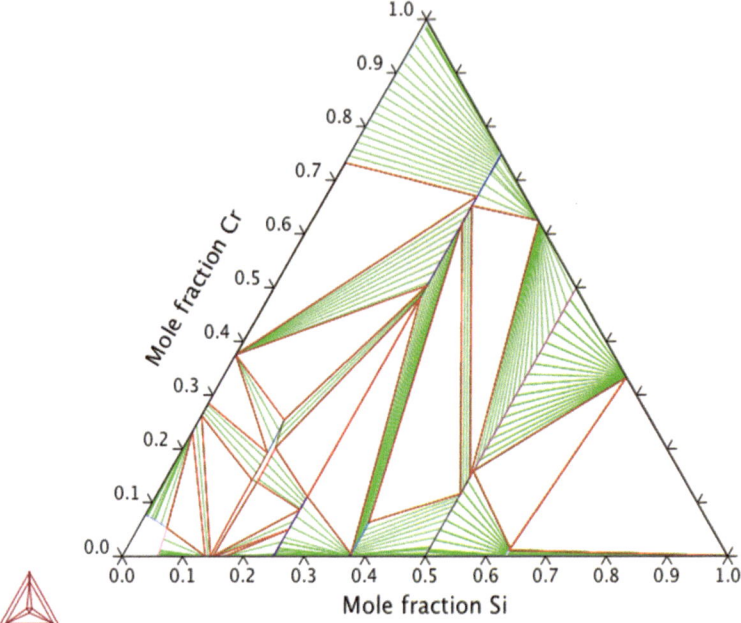

Fig. 5.12 The isothermal section of Cr-Mn-Si at 750 K. Two characteristics are evidenced in this section: the solubility of Cr which is narrow and the section CrSi$_2$-MnSi$_x$

and local equilibrium at the solid/liquid interface which is a general assumption in thermodynamics) [61].

If we go to the Cr-Mn-Si, we obtain the section of Fig. 5.16. Firstly, it has been verified that the amount of Mn in the melt is not so high in order to prevent the precipitation of the B20-MnSi phase which will give finally lower ZT values in the MnSix material. We state the Mn to the 0.97 stoichiometry and then the Si is at 1.74 and Cr (replacing Mn) or Ge (replacing Si) is at 0.03. With those compositions, we obtain Figs. 5.16 and 5.17.

One can observe that in such conditions we have only the HMS (MnSix) phase as pure and the eutectic mixture (MnSix + (SiGe)). Then further heat treatments and/or hot pressing process will reorganize the microstructure as MnSix and (SiGe) solid solution.

Regarding such materials, a second ternary is presented below. Its interest is to understand the solubility of Ge in MnSix and the equilibria between MnSix and (Si,Ge) phase.

In the case where it is necessary to build a database with more than three components, we use the procedure described in Fig. 4.6. It was applied in the system Cr-Ge-Mn-Si for the understanding of the thermoelectric behaviour of the HMS material. Nethertheless in some cases, it can be used also in other applications as magnetism (Mn, Cr-rich alloys) and/or mechanical behaviour. Figure 5.18 represents

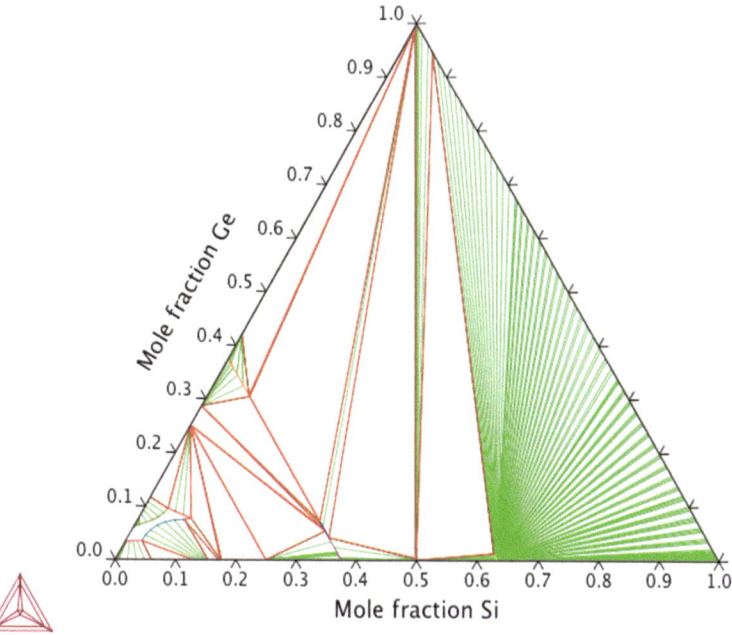

Fig. 5.13 The isothermal section of Ge-Mn-Si at 750 K. Two characteristics are evidenced in this section: the solubility range of Cr which is narrow and the section $CrSi_2$-$MnSi_x$

the deployment in one plane of the tetrahedron based on the four components. This can be done at all possible temperatures.

The phase diagrams as a base of materials are essential for understanding the relationships between composition, structure and properties; this is especially true for complex materials which have variable compositional range. It is important to keep in mind that alloy theory is a key for studying the material properties. For example, the mechanical properties are very dependent with composition and intrinsic defects. In such case, a capital issue concerns the change of the enthalpy of formation of phases as a function of composition.

If we want to understand the behaviour of alloys, it is necessary to have an idea of the phase equilibria of the material. This is necessary but not sufficient. It is why, concerning the materials based on solid solution phases, it is necessary to have precise information on solid solution organization in order to understand their properties. This is why, in many cases, properties of such materials are dramatically varying with chemical composition. For modelling the Gibbs energy of individual phases, the definition of the reference state (^0G_i) is mandatory. A problem that we did not discussed before concerns the phase stabilities. Why is phase stability important?

Prior to the description of phase stabilities, it is necessary to come back to the roots of CALPHAD method; this methodology concerns, as we have seen in Sect. 5.1, the modelling of individual phases of the system. This approach is significant

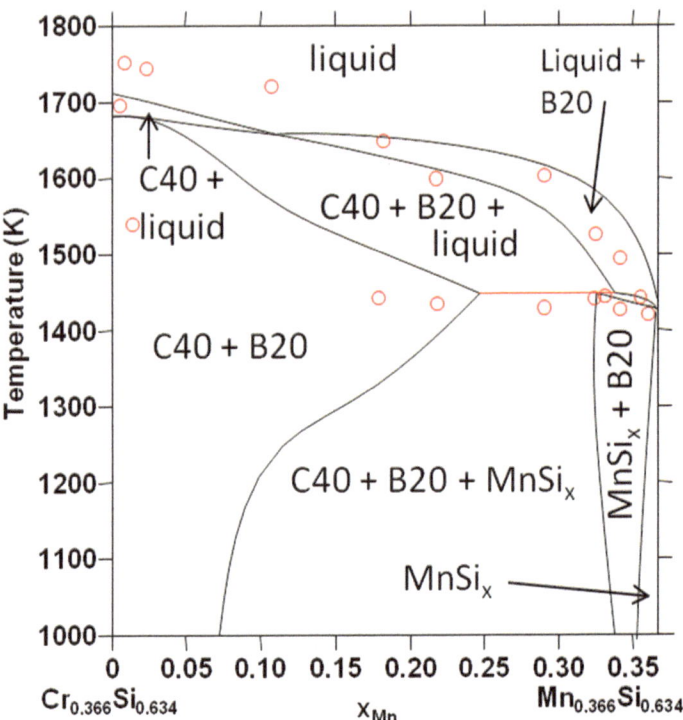

Fig. 5.14 The isopleth section between CrSi$_2$ and MnSi$_x$ showing the different domains of stabilities. The *red open dots* are the experimental values collected from the literature

when it includes the lattice stability of the endmembers of the systems. These endmembers are depending on the difference between stable and nonstable structures of pure elements with respect to the Gibbs free energy. As the phase stability is depending on the stability of elements, the lattice stabilities of elements in different forms should be known with precision. But in some cases (such as in manganese), the stability is each structure is very dependent on the temperature range. Nevertheless, the calculation of phase stability of the whole system (as in Mn-Si system) is necessary; the lattice stabilities of each element in each structure can be obtained by three different ways: experimental (for phases stable in the experimental range), the method of Kaufman and Bernstein [62] which is an extrapolation method [3] and finally the method which is the more accurate presently is first-principle calculation.

The total energies of intermetallic compounds in the systems are calculated by employing the electronic density functional theory (DFT), which uses the pseudopotentials constructed by the projector augmented-wave (PAW) method in the generalized gradient approximation (GGA) for the exchange and correlation energy [64–69]. The calculations are performed for the compounds which are the experimentally observed compounds at their ideal stoichiometry as well as for the

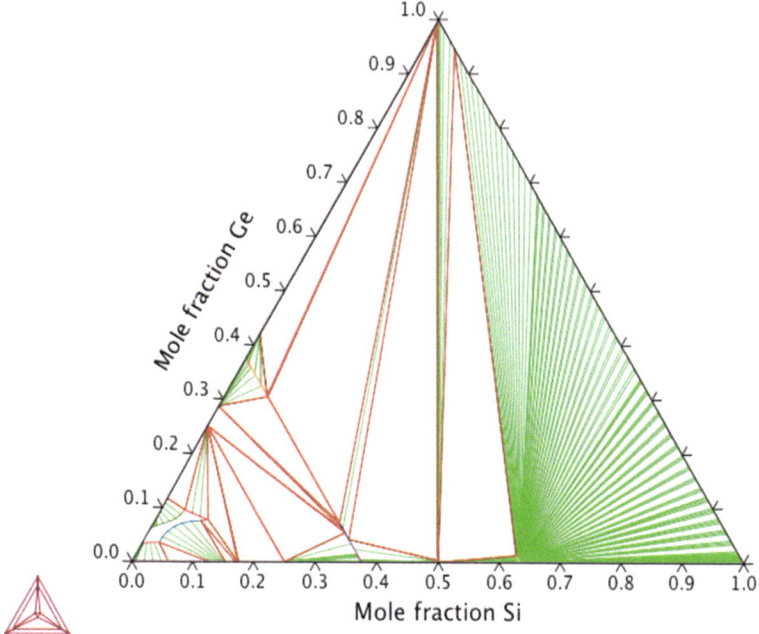

Fig. 5.15 The isothermal section of Ge-Mn-Si at 750 K. It's show s how MnSix can be in equilibrium with the solid solution phase (Si,Ge)

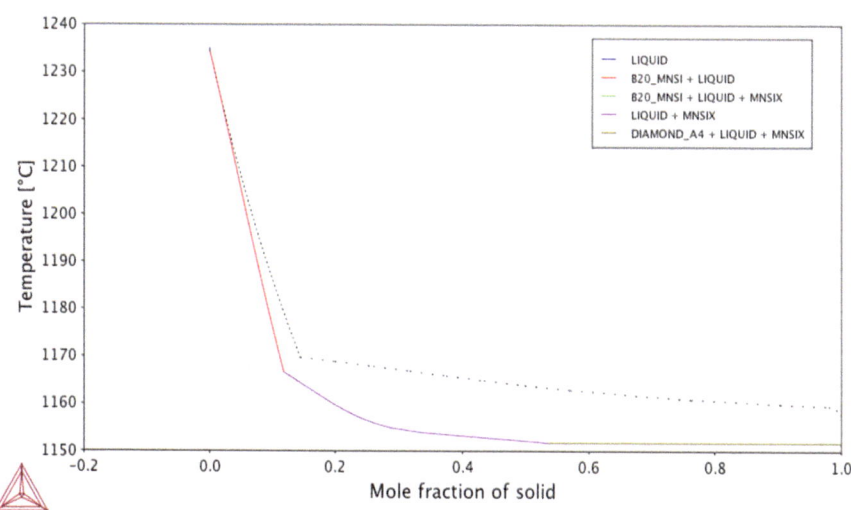

Fig. 5.16 The Scheil simulation of the composition $Mn_{0.97}Cr_{0.03}Si_{1.74}$ giving ZTmax = 0.6 at 850 K [73]

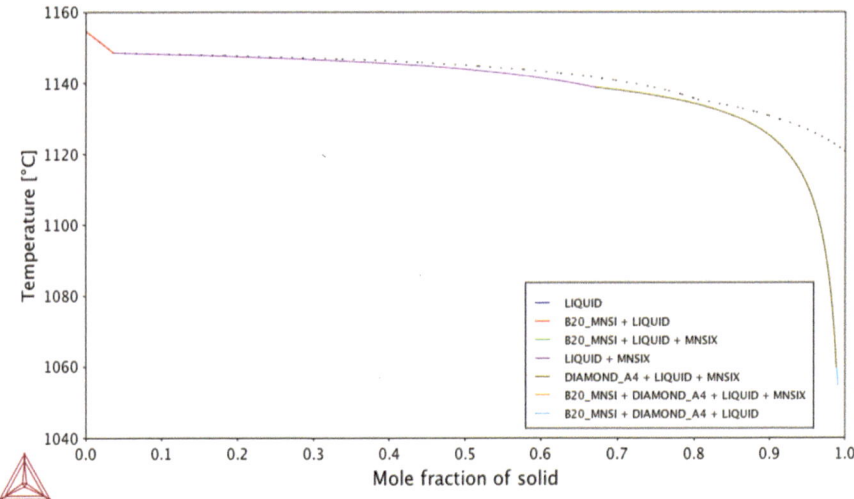

Fig. 5.17 The Scheil simulation of the composition $Mn_{0.97}Ge_{0.03}Si_{1.74}$ giving ZTmax = 0.65 at 850 K [74]

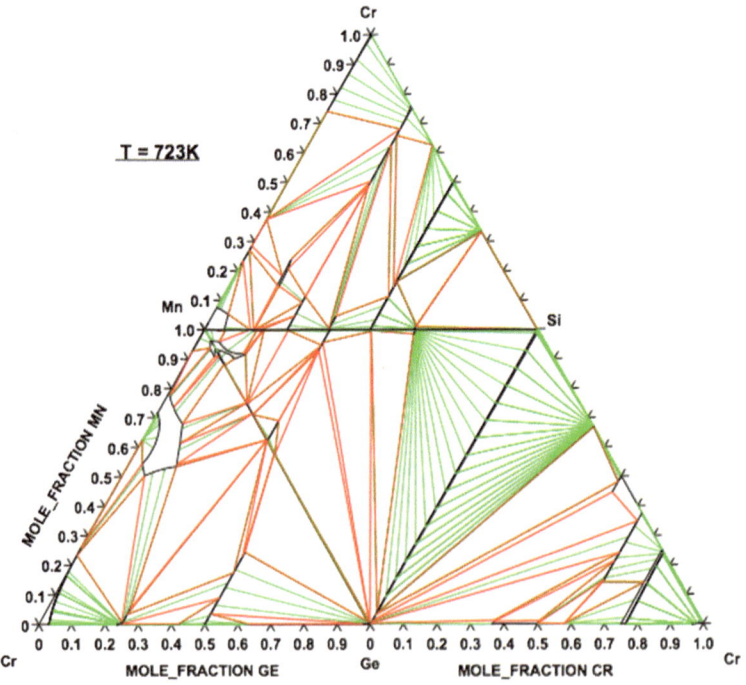

Fig. 5.18 The quaternary system at low temperature showing the equilibria between MnSix and the other phases

structures which can be stable in the systems. Moreover, concerning the nonstoichiometric compounds, the calculation of the energy of formation of defects is mandatory in order to give a good approach of such compounds which are present in many intermetallic phases.

Performing first-principle calculations of the enthalpy of formation of the intermetallic compounds (binary to ternary and other systems) is presently studied in the modern approach of materials. The calculations can also be performed at a given stoichiometry with different crystal structures than those observed experimentally in order to check the possibility of precipitation of metastable phases during alloy processing. Such calculations are also helpful to predict the possible stabilization of structures by ternary additions to a binary system.

Chapter 6
Silicides and Other Thermoelectric Materials

In order to show the place of silicide thermoelectrics among other materials used in thermoelectric generators, it is necessary to consider two points. First, it concerns their sustainability, abundance, inexpensivity and recyclable elements; those materials possess all these positive aspects. Secondly, as one can see on the plot of the figure of merit the dependency of two of the better silicide (HMS and $Mg_2Si_{0.4}Sn_{0.6}$) as a function of temperature, and compared with some good thermoelectric materials working in various temperature ranges, Fig. 4.7. For the base of these figures, we have taken the dependencies from the Ref. [58]. As one can see, magnesium silicide-based thermoelectric is the best material in the temperature range 600–850 K in comparison with the other thermoelectrics of n-type.

© The Author(s) 2017
J.-C. Tedenac, *Multicomponent Silicides for Thermoelectric Materials*,
SpringerBriefs in Materials, DOI 10.1007/978-3-319-58268-9_6

Chapter 7
Conclusion: Applications and Opportunities

This monograph described some properties of silicide-based materials for thermoelectricity. These materials were shown as promising for many applications using thermoelectric effect. Indeed, some materials show high thermoelectric figure of merit ($ZT \geq 1$) and in some cases higher than other well-known compounds.

Efforts worldwide to find available thermoelectric (TE) systems were intensified during the last two decades. The growing needs for alternative energy sources demand new and better materials for energy generation, storage and conversion. Thermoelectric materials (TM) converting heat into power can be of high-impact value for wide-ranging applications in waste heat converters and direct solar thermal energy converters, as soon as efficient materials emerge. Due to the fact that they are environmentally friendly, the silicide materials are in the first place for covering this demand.

Nevertheless, the efficiency of a solid-state thermoelectric engine primarily depends on the chemical compositions of materials and then of phase stabilities. The improvement of the figure of merit in thermoelectric materials depends on multiple factors regarding the microstructures and the physical properties. Then, thermoelectric materials are typically multicomponent systems, and a thermodynamic approach is necessary to study such systems. Also the processes involved in the fabrication of a thermoelectric component lead to other problems which can be solved by thermodynamics:

(a) Solidification processes cause segregations in single phases and, depending on the composition of the melt, eutectic precipitations.
(b) Temperature dependence of the material composition causes point defects leading to imperfect electronic properties.
(c) Hot-pressing processes cause element diffusion in the grains and at the boundaries leading to non-equilibrium material with properties changing with time and temperature.
(d) Nanostructure tailoring entails destabilization and grain coarsening at temperature of use. It is also a question of thermodynamics.

© The Author(s) 2017
J.-C. Tedenac, *Multicomponent Silicides for Thermoelectric Materials*,
SpringerBriefs in Materials, DOI 10.1007/978-3-319-58268-9_7

In the case of such materials, the information on phase transformation, thermodynamic stabilities and process modelling of the new materials presently studied for thermoelectricity are scarce, and it is a pending question.

Adding to these material problems, the relationships between TM and solders and barriers should be studied at the light of thermodynamic tools.

Modern description and phase stabilities of thermoelectric material understanding were described by using thermodynamic description of systems in a general CALPHAD approach. In this approach, the material microstructures after sintering and thermal behaviour can be easily checked (with isothermal section at high temperature) by the comparison with the appropriate isothermal sections. Regarding their potentialities, this method has been applied in some silicide-based systems.

Concerning the system technology, the development of thermoelectric systems based on these materials needs a paradigm shift comparatively to the well-known flat plate design of modules which cannot be considered as exploitable in all applications because of the combination of cost, performance and weight. A number of opportunities can be identified in order to reduce cost and weight and increase performance that could take the current design towards the 75\$/g CO_2 target. The scaling-up potential of processing and manufacturing has to be identified by the considerations of a number of parameters which have to be taken into account up to now for any further technological application which can be considered as exploitable. In fact the situation is very complex. Beyond the ZT values of materials (higher and higher values as shown in literature), one should think that each application needs a special conception and integration systems along the whole chain of material conception and making of systems.Finally, it should be interesting for engineers to understand that for this temperature range application, one should have a multiple choice of materials where they can choose. Based on a full power of 30–40 kW per unit of production, a target cost of around 0.5–1/W can be achieved. However this does not meet the customer expectations in terms of efficiency and cost. Assuming that the performance of an optimized TEG can be achieved at the same cost as the initial system, targeted values of 0.5 / 1 / W for the construction of nextgeneration TEGs are to be considered. The biggest socio-economic impact should be identified as the environmental benefits of the thermoelectric technology process. Implementation and commercialization of this technology in the identified end-user sectors will have a positive environmental impact through improved fuel efficiency and thereby reduced CO_2 emissions. It is important to think that, regarding modern technology, oil consumption for burning is unnecessary nowadays.

The biggest socio-economic impact is identified as the environmental benefits of the thermoelectric technology process. Implementation and commercialization of this technology in the identified end-user sectors will have a positive environmental impact through improved fuel efficiency and thereby reduced CO_2 emissions. Oil is a valuable resource which can be used for more productive purposes than burning unnecessarily. Implementation and commercialization of this technology will enable the civil society to achieve sustainable, competitive and profitable business.

As a concluding remark, it seems clear that a research on all types of silicide materials showing easy route of synthesis is a challenge for going to a huge development of this technology for applications in the range 400–700 K in all applications (generators, sensors, actuators, etc.).

Chapter 8
Databases

The thermodynamic calculations were made by using database files provided by SGTE and our own databases which were published in the journals indicated in the references [70–72]. The MgX database was given to us by Prof. Dr. Rainer Schmid-Fetzer [74], we thank very much.

© The Author(s) 2017
J.-C. Tedenac, *Multicomponent Silicides for Thermoelectric Materials*,
SpringerBriefs in Materials, DOI 10.1007/978-3-319-58268-9_8

References

1. Rowe, D.M., Bandhari, C.M.: Modern Thermoelectrics, pp. 7–25. Reston Publication Company, Reston (1983)
2. Goldsmid, H.J.: General principles and basic considerations. In: Rowe, D.M. (ed.) Thermoelectrics Handbook. Macro to Nano, pp. 19–29-11. CRC/Taylor & Francis, Boca Raton/London/New York (2005)
3. Saunders, N., Miodownik, A.P.: CALPHAD (Calculation of Phase Diagrams), A Comprehensive Guide (Pergamon Materials Series). Pergman, Oxford (9 June 1998)
4. Ioffe, A.: Semiconductors, Thermoelements and Thermoelectric Cooling. Info- search, London (1957)
5. Nikitin, E.N.: Study of temperature dependencies of electrical conductivity and thermal power of silicides, (in Russian). Zhurnal Tekhnicheskoj Fiziki. **28**, 23 (1958)
6. Zaitsev, V.K., Fedorov, M.I., Eremin, I.S., Gurieva, E.A.: Thermoelectrics on the base of solid solutions of Mg2BIV compounds (BIV = Si, Ge, Sn). In: Rowe, D.M. (ed.) Thermoelectrics Handbook. Macro to Nano, pp. 9-1–29-11. CRC/Taylor & Francis, Boca Raton/London/New York (2005)
7. Voronov B.K., Dudkin L.D., Trusova N.N.: The Features of Physical Chemical Structure of Chromium Disilicide (in Russian), Khimicheskaya Svyaz v Poluprovodnikah, p. 291. Nauka i Tekhnika, Minsk (1969)
8. Fedorov, M.I., Zaitsev, V.K.: Thermoelectrics of transition metal silicides. In: Rowe, D.M. (ed.) Thermoelectrics Handbook. Macro to Nano, pp. 31-1–31-19. CRC/Taylor & Francis, Boca/Raton/London/New York (2005)
9. Nagai, H., Takamatsu, T., Iijima, Y., Hayashi, K.: Effects of Nb substitution on thermoelectric properties of CrSi2. J. Alloys Compd. **687**, 37–41 (2016)
10. Karuppaiah, S., Beaudhuin, M., Viennois, R.: Investigation on the thermoelectric properties of nanostructured $Cr_{1-x}Ti_xSi_2$. J. Solid State Chem. **199**, 90–95 (2013)
11. Scherrer, H., Scherrer, S.: Thermoelectric properties of bismuth antimony telluride solid solutions. In: Rowe, D.M. (ed.) Thermoelectrics Handbook. Macro to Nano, pp. 27-1–27-16. CRC/Taylor & Francis, Boca/Raton/London/New York (2005)
12. Gottlieb, U., Sulpice, A., Lambert-Andron, B., Laborde, O.: Magnetic properties of single crystalline Mn_4Si_7. J. Alloys Compd. **361**, 13 (2003)
13. Schwom, O., Preisinger, A., Nowotny, H.: Solid-state synthesis and thermoelectric properties of al-doped MnSi1.75-δ. Wittman A. Monatsch Chem. **95**, 1527 (1964)
14. Knott, H.W., Mueller, M.H., Heaton, I.: The crystal structure of $Mn_{15}Si_{26}$. Acta Crystallogr. **23**, 549 (1967)

© The Author(s) 2017
J.-C. Tedenac, *Multicomponent Silicides for Thermoelectric Materials*,
SpringerBriefs in Materials, DOI 10.1007/978-3-319-58268-9

15. Nowotny, H.: Crystal chemistry of transition element defect silicides and related compounds. In: Eyring, L., O'Keeffe, M. (eds.) The Chemistry of Extended Defects in Nonmetallic Solids. North-Holland Publishing Co., Amsterdam/London (1970)

16. Fredrickson, D.C., Stephen, L., Hoffmann, R., Lin, J.: The Nowotny chimney ladder phases: following the pseudo clue toward an explanation of the 14 electron rule. Inorg. Chem. **43**(20), 6151 (2004)

17. Ivanova, L.D., Baikov, A.A.: Higher manganese silicide based materials. J. Thermoelectricity. **3**, 60 (2009)

18. Zaitsev, V.K., Petrov, Y.U.V., Fedorov, M.I.: The kinetic properties and thermoelectric parameters of partially disordered systems with mobility edge (in Russian). Fizika i tehnika poluprovodnikov. **13**, 1359 (1979)

19. Aoyama, I., Fedorov, M.I., Zaitsev, V.K., Solomkin, F.Y., Eremin, I.S., Samunin, A.Y., Mukoujima, M., Sano, S., Tsuji, T.: Effects of Ge doping on micromorphology of MnSi in MnSi~1.7 and on their thermoelectric transport properties. Jpn. J. Appl. Phys. **44**(12), 8562–8570 (2005)

20. McNeil, D.J.: Iron disilicide as a thermoelectric generator material. Proc. IEEE. **111**(1), 178–182 (1964)

21. Birkholz, U., Gross, E., Stöhrer: In: Rowe, D.M. (ed.) Thermoelectrics Handbook. Macro to Nano, pp. 287–298. CRC/Taylor & Francis, Boca/Raton/London/New York (2005)

22. Ivanov, Y.V., Zaitsev, V.K., Fedorov, M.I.: Contribution of nonequilibrium optical phonons to the Peltier and Seebeck effects in polar semiconductors. Phys. Solid State. **40**(7), 1101–1106 (1998)

23. Fedorov, M.I., Ivanov, Y.V., Vedernikov, M.V., Zaitsev, V.K.: Iron disilicide as a base for new improved thermoelectrics creation. Mater. Res. Soc. Symp. Proc. **545**, 155–160 (1999)

24. Lange, H.: Electronic properties of semiconducting silicides. Phys. Stat. Sol. (b). **201**, 3 (1997)

25. Poutcharovsky, D.J., Parthe, E.: The orthorhombic crystal structure of Ru_2Si_3, Ru_2Ge_3, Os_2Si_3 and Os_2Ge_3. Acta Cryst. **B30**, 2692 (1974)

26. Souptel, D., Behr, G., Ivanenko, L., Vinzelberg, H., Schumann, J.: Floating zone growth and characterization of semiconducting Ru_2Si_3 single crystals. J. Cryst. Growth. **244**, 296–304 (2002)

27. Simkin, B.A., Hayashi, Y., Inui, H.: Directional thermoelectric properties of Ru_2Si_3. Intermetallics. **13**(11), 1225–1232 (2005)

28. Arita, Y., Miyagawa, T., Matsui, T.: Thermoelectric properties of Ru_2Si_3 prepared by FZ and arc melting methods. In: Proceedings ICT98, Seventeenth International Conference on Thermoelectrics, pp. 394–397. IEEE (1998)

29. Vining, C.B.: Extrapolated thermoelectric figure of merit of ruthenium silicide. AIP Conf. Proc. **246**, 338–342 (1992)

30. Arita, Y., Mitsuda, S., Nishi, Y., Matsui, T., Nagasaki, T.: Thermoelectric properties of Rh-doped Ru_2Si_3 prepared by floating zone melting method. J. Nucl. Mater. **294**, 202–205 (2001)

31. Okamoto, N.L., Koyama, T., Kishida, K., Katsushi, T., Inui, H.: Crystal structure and thermoelectric properties of chimney–ladder compounds in the Ru_2Si_3–Mn_4Si_7 pseudobinary system. Acta Mater. **57**, 5036–5045 (2009)

32. Kishida, K., Ishida, A., Koyama, T., Harada, S., Okamoto, N.L., Tanaka, K., et al.: Effects of Nb substitution on thermoelectric properties of CrSi2. Acta Mater. **57**, 2010 (2009)

33. Sakamaki, Y., Kuwabara, K., Jiajun, G., Inui, H., Yamaguchi, M., Yamamoto, A., Obara, H.: Crystal structure and thermoelectric properties of ReSi1.75 based silicides. Mater. Sci. Forum. **426–432**(Npt. 3), 1777–1782 (2003)

34. Oha, M.-W., Gub, J.-J., Inuic, H., Ohd, M.-H., Weea, D.-M.: Evaluation of anisotropic thermoelectric power of ReSi1.75. Phys. B Condens. Matter. **389**(2), 367–371 (2007)

35. Poutcharovsky, D.J., Yvon, K., Parthe, E.: Diffusionless phase transformations of Ru_2Si_3, Ru_2Ge_3 and Ru_2Sn_3 I. Crystal structure investigations. J. Less-Common Met. **40**, 139–144 (1975)

36. Fedorov, M.I., Zaitsev, V.K., Eremin, I.S., Gurieva, E.A., Burkov, A.T., Konstantinov, P.P., Vedernikov, M.V., Samunin, A.Y., Isachenko, G.N., Shabaldin, A.A.: Transport properties of $Mg_2X_{0.4}Sn_{0.6}$ solid solutions (X = Si, Ge) with p-type conductivity. Phys. Solid State. **48**(8), 1486–1490 (2006)
37. Neshpor, V.S., Samsonov, G.V.: Rhenium disilicide as a new refractory semiconductor (in Russian). Fiz. Met. Metalloved. **11**(4), 638–640 (1961)
38. Zaitsev, V.K., Fedorov, M.I., Tarasov, V.I., Adilbekov, A.: Plasma reflection from CoSi and Co1-xNixSi solid solutions. Sov. Phys. Solid State. **19**(6), 996–998 (1977)
39. Geld, P.V., Sidorenko, F.A.: Silicides of Transition Metals of Fourth Period. Metallurgia, Moscow (1991)
40. Kim, S.W., Mishima, Y., Choi, D.C.: Effect of process conditions on the thermoelectric properties of CoSi. Intermetallics. **10**(2), 177–184 (2002)
41. Nikitin, E.N., Bazanov, V.G., Tarasov, V.I.: The thermoelectric properties of solid solution, Mg_2Si-Mg_2Sn. Sov. Phys. Solid State. **3**(12), 2648–2651 (1961)
42. Labotz, R.J., Mason, D.R., O'Kane, D.F.: The thermoelectric properties of mixed crystals $Mg_2GexSi1$-x. J. Electrochem. Soc. **110**(2), 127–134 (1963)
43. Nicolau, M.C.: Material for direct thermoelectric energy conversion with a high figure of merit. In: Proceedings of the International Conference on Thermoelectric Energy Conversion, pp. 59–63. IEEE, Arlington (1976)
44. Fedorov, M.I.: Thermoelectric silicides: past, present and future. J. Thermoelectricity. **2**, 51 (2009)
45. Kim, C.-E., Soon, A., Stampfl, C.: Unraveling the origins of conduction band valley degeneracies in Mg_2Si1 xSnx thermoelectrics. Phys.Chem. **18**, 939 (2016)
46. Kaibe, H.T., Rauscher, L., Fujimoto, S., Kurosawa, T., Kanda, T., Mukoujima, M., Aoyama, I., Ishimabushi, H., Ishida, K., Sano, S.: Development of thermoelectric generating cascade modules using silicide and Bi-Te. In: Proceedings ICT'04, XXIII International Conference on Thermoelectrics, p. 007. IEEE, Australia (2005)
47. Pshenay-Severin, D.A., Fedorov, M.I.: Effect of the band structure on the thermoelectric properties of a semiconductor. Phys. Solid State. **49**(9), 1633–1637 (2007)
48. Yin, K., Xianli, S., Yan, Y., Tang, H., Kanatzidis, M.G., Uher, C., Tang, X.: Morphology modulation of SiC nano-additives for mechanical robust high thermoelectric performance Mg_2Si1 − xSnx/SiC nano-composites. Scr. Mater. **126**, 1–5 (2017)
49. Noda, Y., Kon, H., Furukawa, Y., Nishida, I.A., Masumoto, K.: Temperature dependence of thermoelectric properties of $Mg_2Si_{0.6}Ge_{0.4}$. Mater. Trans. JIM. **33**(9), 851–855 (1992)
50. Zhao, J., Liu, Z., Reid, J., Takarabe, K., Iid, T., Wang, B., Yoshiya, U., Tse, J.S.: Thermoelectric and electrical transport properties of Mg_2Si multi-doped with Sb, Al and Zn. J. Mater. Chem. A. **3**, 19774 (2015)
51. Yin, K., Xianli, S., Yonggao, Y., Yonghui, Y., Qiang, Z., Uher, C., Kanatzidis, M.G., Tang, X.: Optimization of the electronic band structure and the lattice thermal conductivity of solid solutions according to simple calculations: a canonical example of the Mg_2Si1-x-yGexSny ternary solid solution. Chem. Mater. **28–15**, 5538–5554 (2016)
52. Jund, P., Viennois, R., Colinet, C., Hug, G., Fevre, M., Tedenac, J.C.: Lattice stability and formation energies of intrinsic defects in Mg_2Si and Mg_2Ge via first principles simulations. J. Phys. Condens. Matter. **25**(3), 035403 (2013)
53. Gorsse, S., Vivès, S., Bellanger, P., Riou, D., Laversenne, L., Miraglia, S., Clarke, D.R.: Multiscale architectured thermoelectric materials in the $Mg_2(Si,Sn)$ system. Mater. Lett. **166**, 140–144 (2016)
54. Liu, W., Zhang, Q., Yin, K., Uher, C.: High figure of merit and thermoelectric properties of Bi-doped $Mg_2Si_{0.4}Sn_{0.6}$ solid solutions. J. Solid State Chem. **203**(203), 333–339 (2013)
55. Bellanger, P.: Etude de l'influence des paramètres nano et microstructuraux sur les propriétés thermoélectriques des siliciures de magnésium (Mg2(Si,Sn) de type –n. Ph.D. thesis, Université de Bordeaux (2014) (in French)

56. Viennois R., Colinet, Philippe Jund, Jean-Claude Tédenac.: Phase stability of ternary antifluorite type compounds in the quasi-binary systems Mg_2XeMg_2Y (X, Y 1/4 Si, Ge, Sn) via ab-initio calculations. Intermetallics. **31C**, 145–151 (2012)

57. Mao, J., Kim, H.S., Shuai, J., Liu, Z., He, R., Saparamadu, U., Tian, F., Liu, W., Ren, Z.: Thermoelectric properties of materials near the band crossing line in $Mg_2Sn-Mg_2Ge-Mg_2Si$ system. Acta Mater. **103**, 633–642 (2016)

58. El-Genk, M., Saber, H.H., Sakamoto, J., Caillat, T.: Life tests of a skutterudites thermoelectric unicouples (MAR-03). In: Twenty International Conference on Thermoelectrics. Proceedings of ICT'03, pp. 417–420. IEEE, New York (2003)

59. Liu, Z.-K., Wang, Y.: Computational Thermodynamics of Materials. Cambridge University Press, Cambridge (2016)

60. Lukas, H.L., Fries, S.G., Bo, S.: Computational Thermodynamics, The Calphad Method. Cambridge University Press, Cambridge (12 Juillet 2007)

61. Hillert, M.: Phase Equilibria, Phase Diagrams and Phase Transformations: Their Thermodynamic Basis. Cambridge University Press, Cambridge (22 Novembre 2007)

62. Kattner U., Lukas H.L., Petsow G., Gather B., Irle E., Blachnik R.: Effects of Nb substitution on thermoelectric properties of $CrSi2$. Z. fur MetallK. **79**, 32–40 (1988)

63. Kaufman, L., Bernstein, H.: Computer Calculation of Phase Diagrams with Special Reference to Refractory Metals. Academic Press, New York (1970)

64. Kresse G, Furthmüller J.: Efficient iterative schemes for ab initio total-energy calculations using a plane-wave basis set. Comp. Mater. Sci. **6**, 15 (1996); Phys. Rev. B. **54**, 11169 (1996)

65. Blöchl, P.E.: Efficient iterative schemes for ab initio total-energy calculations using a plane-wave basis set. Phys. Rev. B. **50**, 17953 (1994)

66. Kresse, G., Joubert, D.: Efficient iterative schemes for ab initio total-energy calculations using a plane-wave basis set. Phys. Rev. B. **59**, 1758 (1998)

67. Perdew, J.P., Wang, Y.: Efficient iterative schemes for ab initio total-energy calculations using a plane-wave basis set. Phys. Rev. B. **45**, 13244 (1992)

68. Methfessel, M., Paxton, A.T.: Efficient iterative schemes for ab initio total-energy calculations using a plane-wave basis set. Phys. Rev. B. **40**, 3616 (1989)

69. Monkhorst, H.J., Pack, J.D.: Efficient iterative schemes for ab initio total-energy calculations using a plane-wave basis set. Phys. Rev. B. **13**, 5188 (1976)

70. Berche, A., Tédenac, J.C., Jund, P.: Thermodynamic description of the Cr-Mn-Si. Calphad. **55**(2), 181–188 (2016)

71. Berche, A., Tédenac, J.C., Jund, P.: Ab-initio calculations and CALPHAD description of Cr–Ge–Mn and Cr–Ge–Si. Calphad. **49**, 50–57 (2015)

72. Berche, A., Théron-Ruiz, E., Tédenac, J.C., Jund, P.: Thermodynamic study of the Ge–Mn–Si system. J. Alloys Compd. **632**(25), 10–16 (May 2015)

73. Ponnambalam, V., Morelli, D.T., Bhattacharya, S., Tritt, T.M.: The role of simultaneous substitution of Cr and Ru on the thermoelectric properties of defect manganese silicides MnSiTM (1.73o<TM<1.75). J. Alloys Compd. **580**, 598–603 (2013)

74. Kozlov, A., GrÖbner, J., Schmid-Fetzer, R.: Phase formation in MgSnSi and MgSnSiCa. R. JALCOM. **509**, 3326–3337 (2011)

75. Maex, K., van Rossum, M. (eds.): Properties of Metal Silicides. INSPEC, London (1995)

76. Poutcharovsky, D.J., Yvon, K., Parthe, E.: Diffusionless phase transformations of Ru_2Si_3, Ru_2Ge3 and Ru_2Sn3 I. Crystal structure investigations. J. Less-Common Met. **40**, 139–144 (1975)

77. Winkler, U.: Die electrischen eigenschaften der intermetallisher verbindungen Mg_2Si, Mg_2Ge, Mg_2Sn und Mg_2Pb. Helv. Phys. Acta. **28**(7), 633–666 (1955)

Index

© The Author(s) 2017   45
J.-C. Tedenac, *Multicomponent Silicides for Thermoelectric Materials*,
SpringerBriefs in Materials, DOI 10.1007/978-3-319-58268-9